# Mathematical Labyrinths. Pathfinding

# Problem Solving in Mathematics and Beyond

Print ISSN: 2591-7234
Online ISSN: 2591-7242

**Series Editor:** Dr. Alfred S. Posamentier
Distinguished Lecturer
New York City College of Technology - City University of New York

There are countless applications that would be considered problem solving in mathematics and beyond. One could even argue that most of mathematics in one way or another involves solving problems. However, this series is intended to be of interest to the general audience with the sole purpose of demonstrating the power and beauty of mathematics through clever problem-solving experiences.

Each of the books will be aimed at the general audience, which implies that the writing level will be such that it will not engulfed in technical language — rather the language will be simple everyday language so that the focus can remain on the content and not be distracted by unnecessarily sophiscated language. Again, the primary purpose of this series is to approach the topic of mathematics problem-solving in a most appealing and attractive way in order to win more of the general public to appreciate his most important subject rather than to fear it. At the same time we expect that professionals in the scientific community will also find these books attractive, as they will provide many entertaining surprises for the unsuspecting reader.

*Published*

For the complete list of volumes in this series, please visit www.worldscientific.com/series/psmb

**Problem Solving in
Mathematics and Beyond**

Volume **22**

# Mathematical Labyrinths. Pathfinding

Boris Pritsker

**W͞ World Scientific**

NEW JERSEY · LONDON · SINGAPORE · BEIJING · SHANGHAI · HONG KONG · TAIPEI · CHENNAI · TOKYO

*Published by*

World Scientific Publishing Co. Pte. Ltd.

5 Toh Tuck Link, Singapore 596224

*USA office:* 27 Warren Street, Suite 401-402, Hackensack, NJ 07601

*UK office:* 57 Shelton Street, Covent Garden, London WC2H 9HE

Library of Congress Control Number: 2020045922

**British Library Cataloguing-in-Publication Data**
A catalogue record for this book is available from the British Library.

**Problem Solving in Mathematics and Beyond — Vol. 22**
**MATHEMATICAL LABYRINTHS. PATHFINDING**

ISBN 978-981-122-823-0 (hardcover)
ISBN 978-981-123-007-3 (paperback)
ISBN 978-981-122-824-7 (ebook for institutions)
ISBN 978-981-122-825-4 (ebook for individuals)

For any available supplementary material, please visit
https://www.worldscientific.com/worldscibooks/10.1142/12040#t=suppl

Desk Editors: Balamurugan Rajendran/Rok Ting Tan

Typeset by Stallion Press
Email: enquiries@stallionpress.com

In memory of my beloved parents, Polina and Samuil.

To my family: my wife Irina, my younger son Bryan, my older son Aleksandr, his wife Jess, and their daughter, my adorable granddaughter Liana.

Special gratitude goes to Bryan, who despite the intensive curriculum in the sophomore year of his math major in college, always found the time to help me with editing the book and sharing his thoughts and ideas on how to improve its presentation.

# Preface

*Pure mathematics is, in its way, the poetry of logical ideas.*

— Albert Einstein

In everyday life, we all face various challenges and solve various problems. From simple ones to serious projects, and even life-changing decisions, those problems define our life. The internet completely changed the way we live today. An enormous volume of information is available in a matter of seconds. The ability to absorb, analyze the information, and find the optimal answer becomes more and more important. Sometimes, while working under time pressure, it becomes critical to select from the available information the most applicable and essential data to use in getting the desired outcome. No less important is the ability to correctly state, formulate, and

determine the questions regarding a specific issue, to understand
what the ultimate goal is to achieve.

Working as a high school mathematics teacher, I've heard quite
a few times the question-argument "Please, don't tell me I will ever
need those trigonometric functions or properties of a parallelogram in
my life if I am not going to be in any engineering or technical field".
You may substitute words "trigonometric functions and properties
of parallelogram" for any other mathematical terms; the meaning of
the argument will remain the same. Why do we need to bother going
through difficult math subjects if we do not intend to use mathematics in any way in our future profession? Indeed, most likely, the future
literature professor, biology researcher, or archeologist will not need
a deep knowledge of mathematical principals, theorems, and formulas. However, the logical thinking, creativity, imagination, ability to
analyze a problem's conditions and questions asked are invaluable
in any field, no matter what one does for a living. That's what you
get from studying mathematics and learning how to solve problems!
And, the amazing thing is that even the biggest students-skeptics
express interest as soon as you pose some non-trivial logical problems; suggest finding the optimal decision in a tricky situation or
explain a paradox or a sophism involving geometrical constructions
or some algebraic calculations.

I love basketball; it is one of my favorite sports to play and to
watch. My favorite player is the retired great Lithuanian center
Arvydas Sabonis. I was lucky enough to watch him play throughout his whole career, and not just the several final years in the NBA,
when he was playing after receiving multiple dramatic injuries and
lost his speed and athleticism. One of the professional basketball
commentators once called him an amazing combination of Kareem
Abdul Jabar, Magic Johnson, Larry Bird, and Pete Maravich in a 7ft
3in. body! It is no wonder that I am now following his son Domantas
as he makes strides playing basketball in the NBA.

He played his first year for the Oklahoma Thunder. Even though
he is best when he is playing at the center position close to the basket, he was forced by the coach to attempt long-range three pointers
and stay away from the basket. It was a disastrous year. Next season,
Domantas was traded to Indiana Pacers. Indiana's coach utilized his
great rebounding and post-up game, and allowed him to play his natural center position. No more multiple long-distance shots, no more

useless standing away from the basket. He became one of the best and the most consistent players on the team. In 2020, he was even selected as an NBA All-Star. What a great example of how having the same primary conditions, the wrong coach's analysis resulted in a player's failure, while the proper decision of the second team's coach led to the player's game blossoming. One team just lost a great potential leader, and the other one immediately became a play-off contender. How far is this sports-related example of a problem-solving issue from mathematical disciplines? As a matter of fact, conceptually, it is all about solving a problem and finding the right result. This book is about solving non-standard problems, finding the best path to a solution depending on the information given, asking and answering the right questions, and analyzing and comparing different approaches to problem-solving.

I remember from my early childhood how I was captivated by logic puzzles, brainteasers, and riddles from brilliant books written by some of my favorite math authors, Martin Gardner, Boris Kardemsky, and Yakov Perelman. Solving problems in those books, first, with the help of my parents and then by myself, was such a joy! Then, with the years past, while studying mathematical reasoning in my third year in college, I was assigned a project about strategies used by mathematicians to solve problems. Out of the many books I went through, while working on this project, George Polya's *How to Solve It* was a real standout. The book was amazing in drilling to me the overall problem-solving process and it inspired me to self-reflect on how one should approach hard problems.

What does it take to become a successful problem solver? What is the most efficient way to incorporate one's mathematical background and knowledge to tackle challenging puzzles? What does it mean to think outside the box? How can one approach a non-recognizable problem and where does one even start to plan for a solution? Is it advantageous to decompose the problem and examine more accessible problems, introduce auxiliary elements, or consider supplementary problems? These and other similar questions are often posed by kids and students, and unfortunately there is no straightforward answer that satisfies everybody. Good students are usually capable of identifying problems belonging to a specific family, as for instance, the motion or mixture problems, and apply the solving techniques that are best utilized for the specific case. But, sometimes, it becomes stressful if they face a problem unseen before, which they can't relate to any of the known families that are solvable by conventional methods. Even when a difficult problem is solved and well explained, there are still remaining questions: How did you figure out this way of solving the problem? Why did you apply such a method? How did you know that these specific manipulations will lead to the desired result? Why did you introduce this auxiliary element or perform such a construction?

We tried to address these issues and questions while discussing the tactics and strategies in problem-solving by examining different types of unorthodox mathematical problems. From old classic riddles and fun logic puzzles to complicated Olympiad-type math problems, the book offers a selection of more than 200 interesting non-routing problems that require insight, ingenuity, and diligence. We concentrate on a practical approach to a problem-solving process discussing how to examine a problem, thoroughly analyze its conditions and question asked, dissecting it as much as possible to decide what the possible approaches to its solution can be, and finding the proper algorithm or plan to execute such a solution. We also suggest looking for alternative solutions to the same problem, comparing them, and identifying the optimal and the most elegant way to get the result, making interesting conclusions and generalizations from those comparisons. In some cases, the solution of a problem enables us to propose new problems originated as an extension of the considered problem, a converse problem, or an analogy. By "inventing" new problems, we pose new challenges, and gain greater perspectives and valuable insights.

In the first few chapters of the book, we examine different approaches to analyzing a problem. We show hints of how to identify the most essential of the given conditions and attributes, and how to utilize them to simplify a problem's solution. We demonstrate that often a hint to a problem's solution is hidden in the information that is given or even in the way the problem is presented. We concentrate on analyzing the question asked in a problem and show how important it is to clearly understand what is demanded and what must be determined. Sometimes, the key to a solution is in a question asked, and it is critical to answer only what is asked, not what you think is asked, because misunderstanding a problem's question can lead to you solving an entirely different problem. Even though we consider different approaches in separate chapters, the studied tactics and strategies give the best results only when applied in a combination.

In the last few chapters of the book, we face challenging non-traditional problems practicing the suggested first-part analyses. Two chapters are devoted to non-standard geometrical problems not typically covered in high school geometry curriculum. They include constructions with inaccessible elements and study of non-conventional problem-solving techniques applying isometric and homothetic plane transformations. In spite of the problems' non-orthodox character, their solutions require no more than a basic knowledge received during geometry courses. Constructions with inaccessible elements and applications of plane transformations are appealing not merely because of their practical nature but also because the suggested techniques present powerful and rewarding tools in tackling difficult challenges. We also illustrate the usefulness in problem-solving development of playing mathematical and logical games, deciphering mathematical sophisms, interpreting mathematical paradoxes, and inventing new problems.

While working on my previous two books *Geometrical Kaleidoscope*, Dover Publications, 2017, and *The Equations World*, Dover Publications, 2019, my main goal was to illustrate multiple methods and techniques helpful in problems' solutions. I took a practical approach in establishing the links and connections from theoretical issues to their applications in tackling specific problems, while emphasizing the importance of identifying various families of related problems. Recognizing the common features and traits of so-called relatives of the same family gives the ability to apply common techniques for their solutions. I believe it is important to be able to recognize, when possible, the type of a problem and accordingly apply the learned applications and methods for building a problem-solving process. It should simplify it and make it manageable. Both books have been well recognized and accepted by the readers. From readers' reviews, it is evident that the books have been found to be interesting and appealing. We will be referring to the methods and techniques covered in both books here. We will also show a different approaches to several interesting problems discussed previously by finding important links and connections among them. It will not be a repetitious process, but rather discovering the new angles and insights revealing the beauty and elegance of mathematics.

*Mathematical Labyrinths. Pathfinding* is not structured in the same way as the two mentioned books. The problems are selected so that working on each one should teach you something; most of them do not yield to the standard methods. In most cases, we don't depict the problems in sequences, where the problems in a sequence build on preceding problems. The emphasis is placed on problems' solution analyses:

- What is a problem about?
- What is given and how should it be utilized in the most efficient way to solve a problem?
- What is asked? Do we correctly understand a problem's question or what has to be proved?
- How to make a plan for a solution?
- If we don't recognize a problem as belonging to some family of similar problems, how should we approach it?
- What is interesting about the given problem?
- Can we apply the results derived from a solution in more general cases?

Most problems in this book draw upon topics covered in elementary and intermediate algebra and geometry courses; even instances of more complicated Olympiad-type problems require no more than a high school mathematical education. The final chapter offers a collection of "Eureka"-type, out of the ordinary, fun problems that require some bright idea and insight. These intriguing and thought-provoking brainteasers and logic puzzles should be enjoyable by the audience of almost any age group, from 6-year-old children to 80-year-old (and older!) adults.

We see the main goals of this book in stimulating an interest in problem-solving and mathematical thinking development, reinforcing originality, inventiveness, and creativity through analyzing unorthodox problems, finding the right path from labyrinths of non-conventional math challenges, and demystifying the fear of mathematics.

# About the Author

**Boris Pritsker** was born in Kiev, Ukraine. After graduating with a gold medal from secondary school, he studied mathematics at Kiev State Pedagogical University, which he graduated summa cum laude with US equivalent of B.S. degree in math/education. Pritsker worked as a math teacher in high schools including a special math-oriented school for gifted students with advanced programs in algebra, geometry, trigonometry, and calculus. He developed educational seminars for studying classic math problems encouraging students to look for alternative solutions. He also trained school math teams for participation in math competitions; several students from these math teams won the prestigious local and regional math Olympiad contests.

After immigrating to the United States, Pritsker pursued an MBA degree in accountancy. He graduated with an MBA degree magna cum laude from the Graduate school of Baruch College, City University of New York. Pritsker is a licensed CPA in New York State. For the past two decades he has been employed by Marks Paneth LLP, a New York City accounting and consulting firm, where he is currently a director.

Never leaving behind his curiosity and devotion to mathematical education development and research, through the years Pritsker has been exploring various challenging and interesting topics in Euclidean geometry, algebra, and pre-calculus, which he offered for publication in math educational journals. He has published problems and articles in the Soviet Union, USA, Singapore, and Australian magazines such as *Matematika v Shkole* (in Russian), (Mathematics in School), *NY State Mathematics Teachers' Journal, Quantum, New England Mathematics Journal, Mathematics and Informatics, Journal of Recreational Mathematics, Mathematics Teacher, and Mathematics Competitions* (journal of the World Federation of National Mathematics Competitions).

Pritsker authored the books *Geometrical Kaleidoscope*, Dover Publications, 2017 and *The Equations World*, Dover Publications, 2019. Both books have been featured with positive reviews in the Mathematical Association of America website. *Geometrical Kaleidoscope* was assigned a BLL rating and the MAA Basic Library List Committee suggested that undergraduate mathematics libraries consider the book for acquisition.

# Contents

# Chapter 1

# Entering the Labyrinth

When studying mathematics in high school, students solve a wide variety of problems. Solving the same problems, some students show better results than others. They succeed in gaining a deep knowledge of problem-solving techniques and methods, and they don't give up when facing an interesting challenge, they try to overcome it. The others, less successful students, express frustration and even intimidation when seeing unrecognizable problem. They feel confused and usually have no idea how to approach a problem. One of the reasons for this disparity is that some students get into the problem-solving process and are actively involved in it. They analyze problems solved, and try to understand and memorize the techniques used in the solution, while the less successful students thoughtlessly rush to finish the problem; they just try to apply some algorithm without acquiring conceptual understanding of what they are doing and why it is applicable in a specific case. These students have a very vague or even the wrong idea about the whole problem-solving process. When I was teaching mathematics, I heard on quite a few occasions how kids would compare their discouragement to that of being lost in a labyrinth with no way out. One of the goals of this book is to explain important basic fundamentals in the problem-solving process through demonstration of their applications to interesting non-standard problems. Dealing with out-of-the-ordinary problems, general problem-solving strategies often fail or do not deliver the best results. That's why we will cover many artificial tricks and provide useful hints for tackling these problems. We will show how to navigate some "maps" and problem solution plans helpful in finding the way out from arduous mathematical labyrinths.

From psychological studies, it is known that the thinking process is conducted through the following essential aspects: the identification of specific facts or events and breaking a subject into smaller parts to better understand them (analysis), establishing links among them (synthesis), and obtaining a general statement or concept through inference from specific cases (generalization). Analysis, synthesis, and generalization are the "ferments" which digest the products of our outside interactions, impressions, and accumulated knowledge. Generally speaking, to solve a mathematical problem means to discover some sequence of general mathematical terms (definitions, axioms, theorems, laws, rules, formulas) applying which to a problem's conditions we get to the desired result — what was asked to determine or required to prove. As we read the problem the first time, the most important thing is to understand what the problem is about, what the given conditions are, what has to be found or evaluated, and how to link those given attributes to the desired unknown object. So, the analysis is supposed to be our very first step in a problem-solving process. Sometimes, this has to be "documented" somehow. For example, when solving a geometrical problem, it is a good idea to organize and briefly rewrite a problem with mathematical symbols. A picture, diagram, table, or graph should be helpful in making the second step in a solution process, in designing the plan for a solution. If we succeed in conceiving the plan for a solution, its execution will be the next step. As the solution is completed, it is important to verify if it is indeed correct; we need to check the solutions. In many problems, specifically, for example, geometrical construction problems, we also need to investigate how many solutions a problem may have depending on the specific circumstances and clarify the conditions when it has no solutions. Finally, after making all of the above steps, we make a conclusion and write the answer. In many cases, it is very useful to explore alternative solutions, make comparisons among them, decide which one is the most efficient, and research if it is possible to get any interesting outcomes and generalizations from the solved problem.

George Polya's book *How To Solve It* outlines four stages of problem-solving: understanding the problem, devising a plan, carrying out the plan, and examining the result. Tackling non-standard problems, we may wish to highlight several intermediate steps in each of these stages.

So, tentatively we can present a problem-solving process as consisting of the following:

1. Analysis of a problem's conditions and question asked.
2. Sketching the given information in math terms, if possible. Using supplementary diagrams, graphs, tables, etc., if needed.
3. Conceiving a plan for a solution.
4. Carrying out of plan, making actual steps in a problem-solving process.
5. Verifying and checking the results.
6. Exploring how many solutions a problem might have depending on the conditions given.
7. Formulating an answer.
8. Analysis of a problem's solution (finding alternative solutions, generalizations, conclusions).

Before we illustrate the above scheme on a comprehensive example, it worth mentioning that in many word problems, the solution is simplified by translating a problem into algebra terms and introducing variables.

**Problem 1.** There are chickens and rabbits among domestic animals that live on the farm. They have 50 heads and 140 legs. How many chickens and how many rabbits are there on the farm?

**Solution.**

**(1) Analysis of a problem.** We know that each chicken has two legs and each rabbit has four legs. Clearly, we have to use these facts to establish the connections between the given elements. The goal is to determine the number of chickens and rabbits on the farm. Introducing two variables will allow us to interpret the problem's conditions in algebra terms.

**(2) Sketching a problem's conditions in math terms.** Let $x$ be the number of chickens and $y$ be the number of rabbits that live on the farm. They have a total of 50 heads. So, we can translate this condition into equation $x + y = 50$. We also know that they have 140 legs. Since each chicken has two legs and each rabbit has four legs we obtain the second equation $2x + 4y = 140$.

**(3) Conceiving a plan for a solution.** We are interested in finding such values of both variables that satisfy each equation at the same time. Therefore, we have to solve the following system of linear equations:

$$\begin{cases} x + y = 50, \\ 2x + 4y = 140. \end{cases}$$

**(4) Carrying out of plan.** Multiplying the first equation by $-2$ and adding to the second equation yields $2y = 40$, from which we find that $y = 20$. Substituting this into any equation gives $x = 30$.

**(5) Verifying and checking the results.** It is very easy to verify the accuracy of the obtained results: $20 + 30 = 50$ and $2 \cdot 30 + 4 \cdot 20 = 60 + 80 = 140$. We indeed have 50 heads and 140 legs in total among the domestic animals, as it is indicated in the problem.

**(6) Exploring how many solutions a problem has.** Our solution was reduced to solving the system of two linear equations with two variables. Obviously, under the given conditions, there are no other solutions than the pair of numbers 30 and 20.

Perhaps, it is worth it to recall here that generally, if you come across a problem, the solution of which requires a system of linear equations, then the graphical solution of such a system in the $XY$-plane will be the point of intersection of the straight lines representing the graphs of the equations in the system. For the system of two linear equations, there will be one solution, if the lines intersect.

**(7) Answer:** There are 30 chickens and 20 rabbits on the farm.

**(8)** This step is not relevant here, as the arithmetic way of solving this problem is not very appealing compared to the clear and easy algebraic solution.

Before you begin to solve a problem, you have to clearly understand what the problem is about. There are some interesting problems that after reading and analyzing may still leave one confused.

The next problem presents a good example of a challenge usually deemed as non-solvable due to the lack of the information provided. The best strategy when tackling such a problem is to translate its data into algebra, set up an equation or equations linking the given conditions (introduce as many variables as needed), and follow the path step by step working with the equations. It will bring you to some point where you should be able to see the light at the end of the labyrinth and most likely be able to successfully arrive at the desired result.

**Problem 2.** A tourist spent 5 hours sightseeing the center of the city. First, he was walking horizontally at the speed of 4 m/h (miles/per hour), then he went up the hill at the speed of 3 m/h, and returned back at the speed of 6 m/h walking down the hill. Finally, he got back walking through the same horizontal route at the speed of 4 m/h to the point where he started his journey. Find the total walking distance covered by the tourist.

**Solution.** It is interesting to note that for the best results, steps 1 and 2 in this case should be considered together.

**(1–2) Analysis of a problem on a diagram depicting its conditions.** Assume the tourist started walking at point $A$, came to $B$, went up the hill to $C$, returned back downhill to $B$, and finally got to $A$ where he started his trip. This problem clearly belongs to a family of motion problems all of which are solved in one way or another by utilizing the relationship "Distance = speed × time".

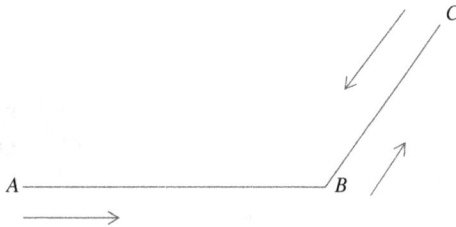

We know the time spent for the whole tour and his speed on each section of the journey, $AB$, $BC$, $CB$, and $BA$. It's not clear, though, how can we calculate the total covered distance without knowing the time spent on each walking segment. If the time would be given, then using the speed on each walking segment, we can find the distances covered and just add them.

**(3) Conceiving a plan for a solution.** Let's introduce a few variables and see how we can use them to link the given conditions into an equation. Let $x$ be the total distance covered on the trip, $x = AB + BC + CB + BA$, and $y$ be the length of the hill, $y = BC = CB$. It is given that the tourist walked for 5 hours. So, if we express the time spent on each of the segments using $x$ and $y$, adding those times should be equal to 5 hours and we will get the equation connecting all the conditions given in the problem. It looks like the original thought of working with distances to get the whole distance is not a good idea; setting up the equation through time periods spent during the trip seems more effective.

Since $AB = BA$, and the tourist was walking at the speed of 4 m/h on that segment on his tour from $A$ to $B$ and returning back from $B$ to $A$, the time spent on the horizontal part of his tour equals $\frac{\frac{x}{2} - y}{4} \cdot 2 = \frac{x - 2y}{4}$ hours. Going uphill, he spent $\frac{y}{3}$ hours and walking downhill he spent $\frac{y}{6}$ hours. Therefore, we see that $\frac{x-2y}{4} + \frac{y}{3} + \frac{y}{6} = 5$.

**(4) Carrying out of plan.** We obtained a linear equation with two variables. Even though it looks confusing at first glance, it makes sense to simplify it as much as possible and see if this would shed light on taking further steps in the solution process.

Finding the common denominator and making simplifications on the left-hand side gives

$$\frac{x - 2y}{4} + \frac{y}{3} + \frac{y}{6} = \frac{3x - 6y + 4y + 2y}{12} = \frac{3x}{12} = \frac{x}{4}.$$

As we can see, the second variable $y$ canceled out, and we arrived at the simplified equation $\frac{x}{4} = 5$, from which $x = 20$ miles.

**(5–6) Verifying the results and exploring how many solutions a problem has.** We found that the total walking distance was 20 miles. In our nominations, the length of the hill is $y$ miles. Then, the length of the horizontal segment is $20 - 2y$ miles. Let's find the total time spent on each segment and see how much time was spent for the entire tour:
$\frac{20-2y}{4} + \frac{y}{3} + \frac{y}{6} = \frac{60-6y+4y+2y}{12} = \frac{60}{12} = 5$. He spent 5 hours for the entire tour, which agrees with the problem's conditions. The problem is solved correctly, and it has the unique solution.

**(7) Answer:** The total distance covered on the tour was 20 miles.

**(8) Analysis of a problem's solution.** It is noteworthy that the second variable $y$ played the role of an auxiliary element that helped establish relationships between the given elements, and linked them into the equation. It was canceled out during our manipulations on the left-hand side of the equation and did not affect the final outcome.

However, it should not diminish its importance in the solution process. We managed to find the total covered distance without determining the walking distance of each specific segment on the tour.

As we observed some basic steps solving the above two relatively easy problems, let's now demonstrate and discuss each of the outlined steps in the solution process with a more complicated example.

**Problem 3.** A motorized boat covered the distance between port docks $A$ and $B$ in 6 hours. On the way back, going with the same speed but against a current, this boat covered the distance from $B$ to $A$ in 8 hours. In how many hours will a banana boat (not a motorized boat) cover the distance between $A$ and $B$ being carried away by a current?

**Solution.**

**(1) Analysis of a problem.** We are dealing with two objects, a motorized boat and a banana boat, one of which has its own speed, while the second is carried away by a current. Similar to Problem 2, there is not much given in the problem. We do not know the distance between $A$ and $B$, we do not know the speed of a motorized boat, nor do we know the current's speed. The goal is to find the time needed for a banana boat to cover the distance between two docks being carried by a current. We can restate the last statement: the goal is to find the time with which the current covers the distance between the two docks.

**(2) Picturing the problem.** The following diagram helps make a vivid illustration of the problem:

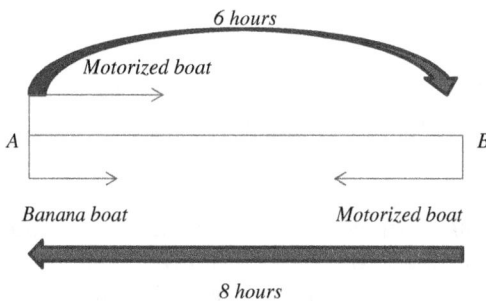

**(3) Designing a plan.** To express the problem's conditions in algebra terms, we introduce several variables. Let $S$ miles be the distance between two docks, $a$ m/h be the current's speed, and $v$ m/h be the motorized boat's speed. Our plan then is in setting up a system of equations connecting the given conditions utilizing these variables.

**(4) Executing the plan.** When the motorized boat is going from $A$ to $B$, it moves at the speed equal to the sum of its own speed and the current's speed, $(v + a)$ m/h. It took 6 hours to get from $A$ to $B$. Therefore, the distance $S$ can be expressed as $S = 6 \cdot (v + a)$ miles, from which

$$v + a = \frac{S}{6}. \tag{1}$$

On its way from $B$ to $A$, the motorized boat's speed is the difference between its own speed and the current's speed, $(v - a)$ m/h. Since the time spent going against the current was 8 hours, the same distance $S$ can now be expressed as $S = 8 \cdot (v - a)$ miles, from which

$$v - a = \frac{S}{8}. \tag{2}$$

The time in question is the time with which the current will cover the distance $S$ while going at speed $a$ m/h. Denoting by $t$ the time needed for a banana boat to go from $A$ to $B$, we can set up one more equation, $t \cdot a = S$.

Our solution is reduced to the following system of equations:

$$\begin{cases} v + a = \dfrac{S}{6}, \\ v - a = \dfrac{S}{8}, \\ t \cdot a = S \end{cases}$$

Subtracting (2) from (1), we obtain that $2a = \frac{S}{6} - \frac{S}{8}$, or equivalently, $a = \frac{S}{48}$. Substituting this for $a$ into the last equation gives $t = \frac{S}{a} = \frac{S}{\frac{S}{48}} = 48$ hours.

**(5) Checking the outcome.** Let's check the result. It suffices to compare the own speed of the motor boat (the speed of the motor boat relative to the water) going with and against the current calculated in two different ways. If we get the same value, the problem is solved correctly. When the boat moves with the current, its speed is the sum of its own speed and the current's speed and equals $\frac{S}{6}$ m/h.

Subtracting from this the current's speed of $\frac{S}{48}$ m/h gives the motor boat's own speed, $\frac{S}{6} - \frac{S}{48} = \frac{7S}{48}$ m/h. On the other hand, when the motor boat moves against the current, its speed is the difference between its own speed and the current's speed and equals $\frac{S}{8}$ m/h. Adding to this the current's speed, gives the motor boat's own speed calculated in a different way, $\frac{S}{8} + \frac{S}{48} = \frac{7S}{48}$. We got the same outcome in both calculations, so we can conclude now that the problem is solved correctly and, indeed, the time needed for a banana boat to go from $A$ to $B$ is 48 hours.

Another way to check the accuracy of our result is to give formulas for $v, a$, and $t$ in terms of $S$ and verify that, with these values, all three equations in our system are satisfied. That would confirm that the claimed solution really meets all the conditions of the problem.

Adding two first equations in the system gives $2v = \frac{S}{6} + \frac{S}{8} = \frac{7S}{24}$, from which $v = \frac{7S}{48}$.

Therefore, $v + a = \frac{7S}{48} + \frac{S}{48} = \frac{8S}{48} = \frac{S}{6}$. We see that $\frac{S}{6} = \frac{S}{6}$, and the first equation is satisfied.

$v - a = \frac{7S}{48} - \frac{S}{48} = \frac{6S}{48} = \frac{S}{8}$. Since $\frac{S}{8} = \frac{S}{8}$, then the second equation is satisfied as well.

Finally, we obtained that $t \cdot a = 48 \cdot \frac{S}{48} = S$. The third equation is satisfied, $S = S$.

**(6) Exploring the solutions.** What is interesting about this problem is that there is not enough information to solve for all four variables. The alternative approach to this particular solution which would address this difficulty would be to set $S = 1$ from the beginning. Setting $S = 1$ is equivalent to choosing the distance between the docks as our unit of distance. This can be justified by noticing that nothing in the problem says that we must use miles as our unit of distance. With this choice of distance units, there is then enough information to solve for the remaining variables. Clearly, the problem has a unique solution.

**(7) Answer:** It takes 48 hours for a banana boat to cover the distance from $A$ to $B$.

**(8) Analysis of the solution.** Let's now go back and discuss our solution. Basically, it was reduced to solving the system of three equations with four variables, not a very conventional system. It may even give an impression that this is not the best way for solving the problem. An alternative approach can be in analyzing the distances covered in 1 hour while going with and against the current. Knowing that the motor boat going with the current covers the distance between the docks in 6 hours and going against the current it covers the same distance in 8 hours, we can conclude that in 1 hour the motor boat covers $\frac{1}{6}$ of the whole distance while going with the current, and it covers $\frac{1}{8}$ of the whole distance while going against the current. The difference $\frac{1}{6} - \frac{1}{8} = \frac{1}{24}$ represents the double distance covered by the banana boat in 1 hour. Hence, the banana boat will cover $\frac{1}{48}$ of the distance between the docks in 1 hour. So, the whole distance will be covered in 48 hours.

In my opinion, this is a more complicated approach than the first one. It is not a clear-cut idea to find the distance covered in 1 hour going with and against the current. The next step for finding the difference between the distances covered in different directions looks even more confusing. Finally, the trickiest part is the last step realizing that the found difference is the *double* distance covered by the banana boat (read — "by the current") in 1 hour.

Another interesting alternative for looking at this problem is by applying the Cartesian coordinate system.

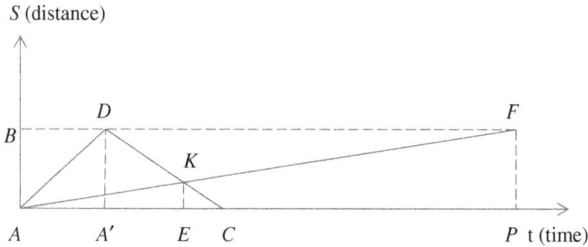

Let's interpret the problem's conditions in graphical form. $AD$ is the graph of the motor boat's movement from $A$ to $B$, $DC$ is the graph of its movement from $B$ to $A$, and $AF$ is the graph of the banana boat's movement (with the current's speed) from $A$ to $B$. We know that $AA' = 6$ and $A'C = 8$. The goal is to find $AP$.

The own speed of the motor boat is the same on the way from $A$ to $B$ and from $B$ to $A$. Therefore, if both, the motor boat and the banana boat, would leave $A$ at the same time, then the motor boat on its way back from $B$ to $A$ will meet the banana boat in the same number of hours as it would spent on its way from $A$ to $B$. Hence, $AA' = A'E = 6$.

Now, we consider two pairs of similar triangles, $\triangle A'DC$ is similar to $\triangle EKC$ and $\triangle KEA$ is similar to $\triangle FPA$. So, we get that $\frac{A'C}{EC} = \frac{DA'}{KE} = \frac{FP}{KE} = \frac{AP}{AE}$. It follows that $\frac{8}{8-6} = \frac{AP}{12}$.

Finally, $AP = \frac{12 \cdot 8}{2} = 48$. We obtained the same result of 48 hours.

Comparing and discussing different approaches, we get a much better understanding of the underlying notions and mental operations involved. It is helpful in consolidating the knowledge obtained during the solution process and developing problem-solving abilities.

In all of the following problems, we will not specifically mention all the steps 1–8 in our solution process assuming readers should be able now to identify these steps, but rather we will concentrate on problem-solving techniques and the mental operations behind them.

There is an important form of analysis that is implemented through synthesis and is very helpful in the solving of many problems. In the thinking process, we often view a subject as involved in new relationships, and as the result, it is seen in playing different roles and possessing some new properties not identified before. In other words, we use the same subject and its multiple properties to establish important links and connections depending on its relationships with other objects. The next problem is a good illustration of this concept.

**Problem 4.** In triangle $ABC$, its angle bisectors $AK$ and $CM$ intersect at $O$. A straight line parallel to $AC$ is drawn through $O$ and it intersects $AB$ and $CB$ at points $D$ and $E$, respectively. Prove that $DE = AD + EC$.

First, let's show how the problem's conditions can be written using the geometrical terms.

**Given:** $\triangle ABC, \angle BAK = \angle CAK, \angle ACM = \angle BCM, AK \cap CM = O$, $O \in DE, DE \parallel AC, D \in AB, E \in BC$.

**To Prove:** $DE = AD + EC$.

**Solution.** Having $DE \parallel AC$, we see that angles $OAC$ and $KOE$ are equal. Indeed, $\angle OAC = \angle AOD$, as alternate interior angles of parallel lines, and $\angle AOD = \angle KOE$ as vertical. So, by transitivity, $\angle OAC = \angle KOE$. Since $AK$ is an angle bisector of angle

$BAC$, then $\angle OAC = \angle OAD$, and by transitivity it follows that $\angle OAD = \angle KOE$. On the other hand, as we mentioned above, angles $AOD$ and $KOE$ are equal as vertical angles, $\angle AOD = \angle KOE$, which implies that $\angle OAD = \angle AOD$. We obtained that triangle $ADO$ is isosceles, with equal sides $AD$ and $DO$, $AD = DO$. In a similar manner, we can show that $OE = EC$. Since $DE = DO + OE$, substituting $DO$ for $AD$ and $OE$ for $EC$, we arrive at the desired conclusion that $DE = AD + EC$.

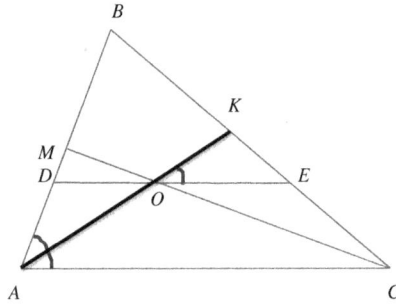

In our solution, we distinguished $AK$ from all other elements (analysis) and it played several different roles:

- $AK$ is an angle bisector of $BAC$,
- $AK$ is a transversal crossing parallel lines $DE$ and $AC$,
- $AK$ is the line having the base of an isosceles triangle $AOD$.

All these different roles of $AK$ are observed in its relationship with other elements in the figure above (synthesis). Depending on their relationships, an analysis through synthesis "redefined" the problem's elements and became instrumental in getting all the parts to be fitted into one logical construction concluding our solution.

If you manage to recognize a problem, meaning you find it as belonging to some related family of problems, you should be able to apply the techniques that are usually helpful for that specific family. A deep knowledge of the subject and past experience from solving various problems are the necessary components in becoming a successful problem solver. However, being well-equipped with a strong mathematical background and formerly acquired knowledge often is not enough when dealing with complex and non-trivial problems.

A good idea is critical in conceiving a plan for a solution. It is impossible to say when it may hit you; the great idea may emerge gradually or it may be a sudden spark, it may even come to you in your sleep. When you are completely lost and don't know where to start, a good piece of advice is to do whatever is doable and see if you get closer to the final goal or at least if you can discover some new ideas enlightening the next steps to take. One thing is for sure, the more you think about the problem, the more different approaches you try, the greater is the chance to overcome the challenge and find your way out of the labyrinth. Sometimes you need to dissect a problem into several pieces, look at it under different angles, introduce some auxiliary elements, as we evidenced in solving Problem 2, or even consider supplementary problems that help reveal the links among given conditions. In many cases, the most brilliant decision becomes the simplest one. The biggest confusion is clarified by a vivid and obvious choice.

Consider this cute logical puzzle:

> *In one of the tourist kiosks in Paris the front page written in English listed the languages in which the suggested city tours for foreign visitors are run during a day. The list has German, Italian, Russian, and many other languages. What language concludes the list?*

Don't you find this to be confusing? What are they talking about? This question does not make any sense! We have no idea how many languages they have on the list and what criteria were used for selecting languages to be included. So, how can we even arrange the unknowns in alphabetical order (this probably is the only logical idea which comes to one's mind)?

Well, let's turn to some supplementary problem and solve another similar "illogical" puzzle:

> *Complete the sentence "If life does not go right, you need to go...".*

Even though this problem also looks like nonsense, the obvious choice for the conclusion becomes clear in a blink of an eye: *go left*, i.e., the whole sentence should be read as *If life does not go right, you need to go left*. Now, it looks fine and even funny.

Returning back to our original problem, we can view it now somewhat differently. We should extend our horizon and try more choices.

And the correct answer, "Finnish", should not look illogical and absurd anymore. Indeed, which language other than Finnish should conclude the list? The final language on the list has to be the one identifying the finish!

I envision an argument from some readers; how can those ambiguously stated problems help in becoming a better math problem solver? Any math problem should be perfectly stated and avoid any ambiguity. Yes, this is correct, and that is the beauty of mathematics. We used the above examples only to illustrate the usefulness of supplementary problems in better understanding and simplifying the problem-solving process. Getting back to math problems, let's examine the following non-trivial equation.

**Problem 5.** Solve the equation:
$$\frac{36}{\sqrt{x-2}} + \frac{4}{\sqrt{y-1}} + 4\sqrt{x-2} + \sqrt{y-1} - 28 = 0.$$

**Solution.** To solve this equation means find all values of two variables $x$ and $y$ satisfying its conditions. Clearly, any methods and techniques applicable for the solution of a single-variable equation should fail in this case. We have to invent something new, design a plan that would simplify the equation and make it manageable.

First of all, note that the permissible domain for the variables is the following:

$x > 2$ (the solutions of inequality $x - 2 > 0$) and
$y > 1$ (the solutions of inequality $y - 1 > 0$).

The suggestion of "do whatever is doable" in this case is reduced to grouping the addends in the left-hand side as

$$\left(\frac{36}{\sqrt{x-2}} + 4\sqrt{x-2}\right) + \left(\frac{4}{\sqrt{y-1}} + \sqrt{y-1}\right) - 28 = 0.$$

We can also factor 4 out of the first parenthesis and rewrite the equation as

$$4\left(\frac{9}{\sqrt{x-2}} + \sqrt{x-2}\right) + \left(\frac{4}{\sqrt{y-1}} + \sqrt{y-1}\right) = 28.$$

Perhaps at this stage this is where one would be stuck. There is not much that can be done if we work with the whole equation, so it should not be a bad idea to "dissect" the problem and look at the equation's components ($\frac{9}{\sqrt{x-2}} + \sqrt{x-2}$) and ($\frac{4}{\sqrt{y-1}} + \sqrt{y-1}$) separately. The next logical step would be to investigate the possible values of $(\frac{9}{\sqrt{x-2}} + \sqrt{x-2})$ and $(\frac{4}{\sqrt{y-1}} + \sqrt{y-1})$, which is the same as investigating the possible values of $\frac{9}{z} + z$ and $\frac{4}{n} + n$ for $z > 0$, $n > 0$. The hint here is to evaluate each of these expressions using the fact that for any positive numbers $t$ and $a$ the following inequality holds true:

$$t + \frac{a}{t} \geq 2\sqrt{a}$$

The question is how one may discover this fact. One could, for example, use a graphing calculator to graph $\frac{9}{z} + z$ and $\frac{4}{n} + n$ for $z > 0$, $n > 0$ to state the hypothesis that each has a minimum value, and then one could try to determine this minimum value.

So, before we proceed with our solution, we need to consider a supplementary problem and prove that, indeed, the inequality $t + \frac{a}{t} \geq 2\sqrt{a}$ holds for any $t$ and $a$ such that $t > 0$ and $a \geq 0$.

Consider and evaluate the difference

$$t + \frac{a}{t} - 2\sqrt{a} = \frac{t^2 + a - 2t\sqrt{a}}{t} = \frac{t^2 - 2t\sqrt{a} + (\sqrt{a})^2}{t} = \frac{(t - \sqrt{a})^2}{t} \geq 0.$$

The last fraction consists of a non-negative nominator, as the square of some number, and a positive denominator, $t > 0$. So, the fraction must be non-negative as well.

It follows that since $t + \frac{a}{t} - 2\sqrt{a} \geq 0$, then indeed, $t + \frac{a}{t} \geq 2\sqrt{a}$, and we can apply this result to evaluate each of the components in the original equation. In our nominations, in the first case, $a = 9$ and $t = \sqrt{x-2}$, therefore,

$$\frac{9}{\sqrt{x-2}} + \sqrt{x-2} \geq 2\sqrt{9} = 2 \cdot 3 = 6.$$

In the second case, $a = 4$ and $t = \sqrt{y-1}$, therefore,

$$\frac{4}{\sqrt{y-1}} + \sqrt{y-1} \geq 2\sqrt{4} = 2 \cdot 2 = 4.$$

So, we see that $4(\frac{9}{\sqrt{x-2}} + \sqrt{x-2}) + (\frac{4}{\sqrt{y-1}} + \sqrt{y-1}) \geq 4 \cdot 6 + 4 = 28$. The equality will be possible only when $4(\frac{9}{\sqrt{x-2}} + \sqrt{x-2}) = 24$ and $\frac{4}{\sqrt{y-1}} + \sqrt{y-1} = 4$, which is attainable when $\sqrt{x-2} = 3$ and $\sqrt{y-1} = 2$ (the minimal possible values based on our analysis). Solving the last two equations gives $x = 11$ and $y = 5$.

Verifying the results, we obtain
$$(\frac{36}{\sqrt{11-2}} + 4\sqrt{11} - 2) + (\frac{4}{\sqrt{5-1}} + \sqrt{5} - 1) - 28 = 24 + 4 - 28 = 0; 0 = 0$$
is true, and we arrive at the conclusion that the equation is solved correctly.

**Answer:** $x = 11$ and $y = 5$.

Analyzing the steps in our solution process and the obtained result, we have a good opportunity to extend this problem and raise questions leading to a few interesting new problems. Once we established the fact that $t + \frac{a}{t} \geq 2\sqrt{a}$ for $t > 0$ and $a \geq 0$, we discovered that it is only for the specific value 28 on the right-hand side of the equation that we get a unique solution. Let's discuss the collection of solutions if the right-hand side had been some other number. For example, if we change 28 to 27, there are no solutions. If we change 28 to 29, instead of concluding that $\frac{4}{\sqrt{y-1}} + \sqrt{y-1}$ must take the value of 4, we conclude that it can have any value between 4 and 5, which leads to the conclusion that $y$ can be any number between 2 and 17. Let's see how this can be justified.

Since we now consider $y$ such that $4 \leq \frac{4}{\sqrt{y-1}} + \sqrt{y-1} \leq 5$, then to find the values of $y$ satisfying this inequality, we will make a substitution $\sqrt{y-1} = u$, $u \geq 0$, and we will solve the inequality $4 \leq \frac{4}{u} + u \leq 5$. In fact, since $\frac{4}{u} + u \geq 4$ for any $u > 0$, we need to consider only $\frac{4}{u} + u \leq 5$. This is equivalent to $u^2 - 5u + 4 \leq 0$ for $u > 0$. Factoring the left-hand side as $(u-1)(u-4) \leq 0$ and solving the last inequality, we get that $1 \leq u \leq 4$.

Going back to our substitution, we see that $1 \leq \sqrt{y-1} \leq 4$ or equivalently,

$$2 \leq y \leq 17.$$

For each value of $y$, we can then solve for corresponding values of $x$. The reader may wish to finish the calculations and find the range for all such values of $x$.

To conclude the discussion of the solution of Problem 5, it merits mentioning that we managed not only to solve a non-routine equation, but derived a useful property $t + \frac{a}{t} \geq 2\sqrt{a}$ for $t > 0$ and $a \geq 0$ that is worth remembering for future explorations. We also evidenced that by slightly modifying this problem's conditions, we came across new interesting challenges.

To appreciate the following two problems, they have to be considered in a combination. The first problem will be decomposed into several small problems, which will enlighten the bigger picture. Emphasis should be put on the importance of deep math knowledge of various topics and the ability to get efficient results utilizing such skills. We will be working backwards putting solved supplementary problems together to arrive at the desired result. The second problem is a cute puzzle which looks extremely difficult and usually is perceived as related to the previous problem. Considering them together adds to this confusion. In reality, it turned out that the problems have almost nothing to do with each other and the solution of the second one requires only a strong analysis and understanding of what it is about; it is one of those problems solving which one just can't appreciate enough the beauty and elegancy of mathematics.

**Problem 6.** Evaluate
$\log \tan 1° + \log \tan 2° + \log \tan 3° + \cdots + \log \tan 89°.$

**Solution.** Having the sum of the logarithms, the natural desire in taking the first step is applying the property that $\log_a m + \log_a n = \log_a(m \cdot n)$, $m > 0$, $n > 0$. We can modify the given expression as

$$\log \tan 1° + \log \tan 2° + \log \tan 3° + \cdots + \log \tan 89°$$
$$= \log(\tan 1° \cdot \tan 2° \cdot \tan 3° \cdot \ldots \cdot \tan 89°).$$

To proceed further, we need to work with the product of tangents of angles from $1°$ to $89°$ and consider the supplementary problem:
Simplify the expression $(\tan 1° \cdot \tan 2° \cdot \tan 3° \cdot \ldots \cdot \tan 89°)$.
To succeed, we need to recall the properties of trigonometric functions that $\cot \alpha = \tan(90° - \alpha)$ for all angles $\alpha$ for which functions $\tan \alpha$ and $\cot \alpha$ are defined.

Observing that

$$\tan 89° = \tan (90° - 1°) = \cot 1°,$$
$$\tan 88° = \tan (90° - 2°) = \cot 2°,$$
$$\tan 87° = \tan (90° - 3°) = \cot 3°,$$

$$\ldots\ldots\ldots\ldots\ldots\ldots\ldots\ldots\ldots\ldots\ldots$$

$$\tan 46° = \tan (90° - 44°) = \cot 44°$$

and having $\tan 45°$ as the mid-term factor in our product, we can rewrite the expression as
$\tan 1° \cdot \tan 2° \cdot \ldots \cdot \tan 89° = \tan 1° \cdot \tan 2° \cdot \ldots \cdot \tan 45° \cdot \ldots \cdot \cot 2° \cdot \cot 1°$.
Grouping equidistant from both end factors and recalling that $\tan 45° = 1$ and $\tan \alpha \cdot \cot \alpha = 1$, the last equality modifies to

$$(\tan 1° \cdot \cot 1°) \cdot (\tan 2° \cdot \cot 2°) \cdot \ldots \cdot (\tan 44° \cdot \cot 44°) \cdot \tan 45°$$

$$= 1 \cdot 1 \cdot \ldots \cdot 1 = 1.$$

So, now, we can return to the original expression and see that

$$\log \tan 1° + \log \tan 2° + \log \tan 3° + \ldots + \log \tan 89° = \log 1 = 0.$$

The problem is solved.

**Problem 7.** Evaluate $\log \tan 1° \cdot \log \tan 2° \cdot \log \tan 3° \cdot \ldots \cdot \log \tan 89°$.

**Solution.** At first glance, it is not clear how to modify this expression and where even to start. But, after looking carefully at what we are dealing with, one should realize that nothing has to be modified or simplified at all! There is the product of logarithms of tangents of all the angles from $1°$ to $89°$. Therefore, one of the factors is $\log \tan 45°$. Since $\tan 45° = 1$, it implies that $\log \tan 45° = \log 1 = 0$. Therefore, the whole product equals to 0.

If solving the first of these two problems we required the knowledge of the properties of logarithms and trigonometric functions, needed some creativity to manipulate the factors in the most efficient way to get to the result, then in the second problem, all we needed was the concentration and proper analysis of the given conditions. The lesson learned — always pay attention to everything that is given in a problem, even a small detail may play a critical role in its solution. The seemingly difficult problem turned out to become a math joke to enjoy!

**Problem 8.** Solve the equation $4\sin \pi x = 4x^2 - 4x + 5$.

**Solution.** First, let's try simplifying the equation a little bit by dividing both sides by 4. We will get $\sin \pi x = x^2 - x + \frac{5}{4}$.

This non-standard equation has a trigonometric expression on its left-hand side and quadratic trinomial on the right-hand side. Most likely, a straightforward approach and conventional techniques are bound to fail here. Clearly, we have to come up with some bright idea that would shed light on the next steps to take. Dealing with non-standard equations, a so-called assessment technique of the ranges of both sides of an equation often proves useful. Having a quadratic trinomial on the right-hand side, we can try to complete a square and see if it would clarify the picture:

$$\sin \pi x = \left(x - \frac{1}{2}\right)^2 + 1.$$

The range of the function sine is all real numbers not exceeding 1 in absolute value. So, the expression on the left-hand side does not exceed 1 in absolute value, i.e., $|\sin \pi x| \leq 1$. On the other hand, the expression on the right-hand side is greater than or equal to 1, $\left(x - \frac{1}{2}\right)^2 + 1 \geq 0 + 1 = 1$. The equality, therefore, is possible only when $x = \frac{1}{2}$, allowing the right-hand side to attain its minimum value of 1 and the left-hand side — its maximum value of 1. There are no other solutions.

Similarly to Problem 5, Problem 8 is a bit of a trick question; the quadratic on the right was chosen very carefully to make a solution possible. By changing 4 on the left-hand side to some other value, we would get a different problem, investigation of which we leave to the ambitious reader.

As we mentioned above, when facing non-standard problems, one cannot underestimate the role of a lucky guess, an assumption based on intuition, and a "sudden" bright idea illuminating the whole picture and making a problem manageable. But, coming off a good idea is not something that is "God given". Even failed attempts and unsuccessful problem-solving plans take you closer to the desired result. Considering different approaches, thinking about your previous steps and evaluating what went wrong, looking for similar or supplementary problems, introducing auxiliary elements, and reviewing your previous analysis are the steps leading to progress in the solution process. You are getting closer to the ultimate goal. Thinking it over and over again, you have a deeper understanding of a problem. Assessing and pondering a veritable smorgasbord of ideas, you reject worthless ones, hoping for the emergence of the right one. To get the proper guess or a breakthrough idea out of multiple thoughts, one needs to assess its suitability and effectiveness for a specific problem. A bright idea is usually based on obtained knowledge and prior experience and it is always a product of your thinking activities.

**Problem 9.** The wisest men of the kingdom are called upon by the king to his court. One of them is to be chosen for the advisor rank and thus they should take part in a contest solving intelligence tests. After several days of competitions, only two winners emerged. The king called both of them to his palace and said, "Tomorrow morning I will write on a paper and hide three digits $a$, $b$, and $c$ which you would need to identify. You will have to tell me any three numbers $x$, $y$, and $z$. I will calculate then the sum $ax + by + cz$ and tell you the result. Knowing this number, whoever guesses the three digits $a$, $b$, and $c$ which I have in hiding, will be the winner". What is a possible strategy to win the competition and solve the king's problem?

**Solution.** This is one of those classic problems where you need to come up with some bright idea; you have to invent your own solution, which may not be unique. Don't be fooled by the king's words, the competitors can't just guess the three digits he had in mind. They need to come up with very specific numbers $x$, $y$, and $z$ (there are no restrictions on what numbers to pick!) allowing them to figure out each of $a$, $b$, and $c$ knowing only the resulting sum of products, $N = ax + by + cz$. Clearly, it's not a guessing game, but rather a challenging problem requiring some unorthodox approach for solving it. This problem might have indefinitely many solutions. To simplify it, we may put one of our numbers, lets' say $x$, to equal 0. Then, the problem will be reduced to the equation with two variables. Indeed, we will know $N$, the resulting sum, and we know our selections for $y$ and $z$, so the equation is simplified to $N = by + cz$. Selecting integer values for $y$ and $z$, we can get a Diophantine equation (an equation with two or more variables whose values are restricted to integers;

Diophantus of Alexandria was the first to study such equations in the 3rd century A.D.) and apply techniques available for solving some types of such equations. We can get solutions for $b$ and $c$. However, we still will have no clue for correctly identifying $a$. So, this is a way nowhere.

Let's start from the beginning and recall the king's demand. He was talking about having in hiding the three specific digits. What do we know about digits? There are only 10 of them, $0, 1, 2, \ldots, 9$, and they are used in composing any other number. So, if we find some method allowing us to preserve them in the final sum $ax + by + cz$ as written on specific spots, it will solve the problem. It means we need to reduce our problem to get the final sum as a three-digit number such that it has the form of $\overline{abc}$. But, the expanded presentation of any three-digit number is $\overline{abc} = 100 \cdot a + 10 \cdot b + 1 \cdot c$. The expanded form is a way to write a number such *that all of the place value components of the number are separated.* When we write a number in expanded form, each digit is broken out and multiplied by its place value, such that the sum of all of the values equals the original number.

The bright idea, we were looking for, is to select our numbers to give to the king as $x = 100$, $y = 10$, and $z = 1$. No matter what the king's original selection of the values for the digits $a$, $b$, and $c$ is, the result of the sum $ax + by + cz$ will be a three-digit number written as $\overline{abc}$. So, we will be able to tell the king each of his digits just by looking at the sum. For example, if the king got $a = 4$, $b = 9$, and $c = 6$, after multiplying $4 \times 100$, $9 \times 10$ and $6 \times 1$, the sum will be 496. This number will reveal to us each of the three hidden digits 4, 9, and 6.

Analyzing the problem's conditions, trying different approaches, overcoming the failed decisions, we get closer to the idea of an expanded three-digit number presentation. This enlightened the idea of selecting $x$, $y$, and $z$ as place values in expanded form. Our bright idea was a rewarding outcome of all the thinking activities, not an unexpected lucky gift or guess.

Getting acquainted with the major steps in a problem-solving process, we will concentrate in the next two chapters on analysis of a problem's conditions and questions asked. We will reveal some thoughts and ideas of taking full advantage of what is given to us in conceiving the pathfinding from tricky math labyrinths. Many of these ideas will be encountered in subsequent chapters as well.

# Chapter 2

# What? How? Why?

We begin with working in logic, leaving strictly mathematical topics for the second part of the chapter. The exercises, including several classic riddles, are selected in such a way that we learn cogitation and not tedious calculations. This chapter collects problems in which the basic difficulty may lie in making the distinction between relevant and irrelevant data or between known and unknown information, sometimes even misleading information. Each problem here has its own identity. The emphasis is on pinpointing such an identity and revealing how the given information dictates the selection of the best strategy for tackling the problem.

When one starts solving a problem, one has to clearly understand *what* the problem is about and *what* data is to be used in contemplating the plan for solution. The next question is *how* to use this data and the given attributes and *how* to link them in a logical chain so they could lead to the desired result. Finally, *why* did we solve the problem this way, can it be applied to other problems, and what is so special and noteworthy about the derived solution? In this chapter, we will try to address all these questions and give some answers while looking for a pathfinding from the following challenging labyrinths.

**Problem 1.** If one worker can pack 10 boxes every 5 minutes, and another can pack 10 boxes every 4 minutes, how many minutes will it take these two workers, working together, to pack 180 boxes?

**Solution.** The key to the problem is to get to a common rate. We have to determine the number of boxes they will pack working together for the same time. It takes 5 minutes for the first worker and 4 minutes for the second worker to pack the same number of boxes, 10 boxes. The least common multiple of 4 and 5 is 20. The first worker will pack 40 boxes in 20 minutes, while the second worker will pack 50 boxes in 20 minutes. So, working together, they will pack 90 boxes in 20 minutes. Clearly, the double of that, 180 boxes, they will pack in 40 minutes.

Let's now take a slightly different view at this problem, and consider the rates at which the two workers pack boxes: the first worker packs 2 boxes/minute, and the second worker packs $\frac{5}{2}$ boxes/minute. We see that together they pack $2 + \frac{5}{2} = \frac{9}{2}$ boxes/minute, and therefore, to pack 180 boxes, it should take $180/(\frac{9}{2}) = 40$ minutes. The approach outlined above should work, but we need to be careful. For example, suppose the question had asked how long it would take them to pack 100 boxes. Using the above method, one may think that the answer is $\frac{100}{\frac{9}{2}} = \frac{200}{9} = 22\frac{2}{9}$ minutes. But, in $22\frac{2}{9}$ minutes, the first worker would pack $22\frac{2}{9} \cdot 2 = 44\frac{4}{9}$ boxes and the second worker would pack $22\frac{2}{9} \cdot \frac{5}{2} = 55\frac{5}{9}$ boxes, giving a total of 99 fully packed boxes and two partially packed boxes, one $\frac{4}{9}$ full and one $\frac{5}{9}$ full. This is not the same as 100 packed boxes. Since we get 99 full boxes, there has to pass some time for one of the workers to finish packing the box he's in the middle of packing. It would take less time for the second worker (at his speed of packing) to finish packing the full box because he has to pack $\frac{4}{9}$ of the box to finish it, while the first worker needs to pack $\frac{5}{9}$ of the unfinished box. It would take $\frac{\frac{4}{9}}{\frac{5}{2}} = \frac{8}{45}$ minutes for the second worker to finish partially packed 56th box. Accordingly, it would take him a total of $22\frac{2}{9} + \frac{8}{45} = 22\frac{2}{5}$ minutes to get 56 fully packed boxes. In $22\frac{2}{5}$ minutes, the first worker would pack $22\frac{2}{5} \cdot 2 = 44\frac{4}{5}$ boxes and adding 56 boxes packed by the second worker for the same time, we get a total of 100 boxes plus a partially filled box. So, the correct answer is $22\frac{2}{5}$ minutes.

The point here is that one has to be extra cautious working with the specific conditions while answering the question even to a conceptually similar problem. It merits also reminding that one should always check one's answer.

In some problems, the language might be confusing and even misleading, as it is, for instance, in the following Problems 2 and 3.

**Problem 2.** Two neighbor-farmers sold apples in the following way:
The first farmer sold 2 apples for $1, the second farmer sold 3 apples for $2. Each one has 30 apples available for sale on this particular day. So, assuming all the apples will be sold, the first farmer expected to get the gross sales proceeds of $15 and the second one expected the gross sales of $20, giving the total sales for both to be $35. In order not to compete with each other and attract more customers, they decided to share the apples and sell them together. They agreed to put five apples in one package and sell each such package for $3. Their rationale was the following: if one of us is selling two apples for $1 and the second farmer is selling three apples for $2, then it has to be fair to sell five apples for $3. This tactic worked well and they sold all the apples by the end of the day for $36. Both farmers were surprised that they ended up getting $1 more than they expected, and they had no idea how to fairly split an extra $1 between them. What was wrong in their calculations?

**Solution.** By making the packages have five apples in each and assigning $3 price per such package, they were selling the apples at a different price per apple versus their individual sales. Indeed, the first farmer had intended to sell apples at $\frac{1}{2}$ per apple and the second farmer had intended to sell apples at $\frac{2}{3}$ per apple, while the selling price of one apple in a package was $\frac{3}{5}$ per apple. Selling the apples separately, the first farmer would make $30 \cdot \frac{1}{2} = \$15$, and the second farmer would make $30 \cdot \frac{2}{3} = \$20$, making the total equal to $35. Combining their efforts and selling apples in packages, they together made $60 \cdot \frac{3}{5} = \$36$. An extra $1 is attributable to a different price per dollar assigned in each package. So, the tricky outcome of the farmers' calculations was clarified by thorough analysis of the known information.

Another way to look at this problem is to compare the cash receipts of each farmer for his contributions to the combined package. The farmers make up their packages of five apples by having the first farmer put two apples in the package and the second put in three apples. Hence, when the package is sold for $3, the first farmer would keep $1 (the price of his two apples) and the second farmer would keep $2. This would work fine for the first 10 packages of apples. But after they sold 10 packages, the second farmer would be out of apples and the first would have 10 apples left. If he put those remaining 10 apples in packages of five and sold each package for $3, he would be selling his last 10 apples for $0.60 each, rather than for $0.50, making an extra $0.10 per apple sold. This explains where the extra $1 comes from.

The next problem presents one of the variations of the famous *Missing dollar riddle* that involves an informal fallacy.

**Problem 3.** Three men in a café ordered a meal of which the total cost was $15. They paid $5 each. The waiter took the money to the café owner, who recognized the three as friends and asked the waiter to return $5 to them. The waiter instead of going to the trouble of equally splitting $5 among the three gave them $1 each and dishonestly pocketed the remaining $2 for himself. As a result, each of the three men paid $4, making the total meal cost $12. Add the $2 pocketed by the waiter and this comes to $14. Where has the other $1 gone from the original $15 paid for the meal?

**Solution.** Similar to the previous problem, we have to clarify what the problem is about and compare the comparable items. The trick here is to realize that the $12 actually paid to the café owner already includes $2 pocketed by the waiter. To add the $2 to the $12 would be to double count it. The mistake is the phrase "add $2 pocketed by the waiter". In fact, one should take $12 paid by the men and subtract the $2 pocketed by the waiter to see that the owner got $10. In other words, we need to see how much the actual cash outflow was. To obtain a sum that totals the original $15, every dollar must be accounted for, regardless of its location. The café owner got $15 - 5 = \$10$. Each of the three men was left with $1 returned to them by the waiter. Finally, the dishonest waiter pocketed $2. The total then is $10 + 3 + 2 = \$15$, the same amount as the originally was spent for the meal. This logical and clear explanation can be called the "accounting model" — counting assets in everybody's possession gives the total funds available for spending before their distribution.

**Problem 4.** In medieval times, one tourist arrived at a hotel and asked the hotel owner to let him stay for seven nights. He said that he had no money, but he can pay for his stay with a seven-ring golden chain, counting one ring per night. The hotel owner agreed but demanded that the payment to be made for each night up-front, and asked that the least possible cuts were made in the chain. How many cuts does the tourist have to make to fulfill the hotel owner's demand and still be able to pay one golden ring per night for his stay?

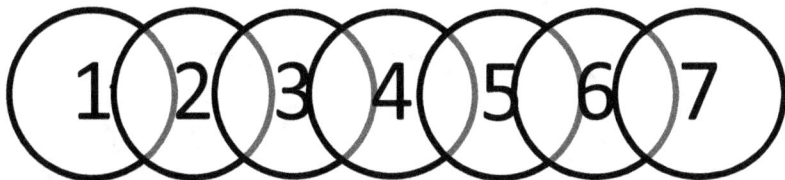

**Solution.** First of all, one has to realize that one cut represents the cut of one ring. Many times, I have heard people say to cut across as a suggested solution. If you do that though, you have seven cuts, not one cut, so it would represent the maximum, not the minimum number of cuts. What I found the most confusing about this problem is that people usually perceive the question as how can one perform the minimum number of cuts while giving up one ring per night, instead of paying one ring per night. The whole idea is that the hotel owner asked the tourist *to pay* him one ring per night, which does not necessarily mean *to physically give* him one ring per night. As soon as you clarify this, the solution becomes manageable; it can be related to an important mathematical principle, binary representation of positive integers. The number of rings in the chain can be expressed as

$$7 = 2^0 + 2^1 + 2^2 = 1 + 2 + 4.$$

So, the pieces of the gold chain we need to work with should have size 1, 2, and 4. Clearly, we can get these pieces by making just one cut of the third ring from either side (the problem has two solutions). By cutting the third ring, for example, we can take it out from the chain having on hands one ring, the third ring, two connected rings, the first and the second, and four connected rings, the fourth, the fifth, the sixth, and the seventh. Now, the question is how the tourist should be able to fulfill the hotel owner's demand and *pay* one ring per night.

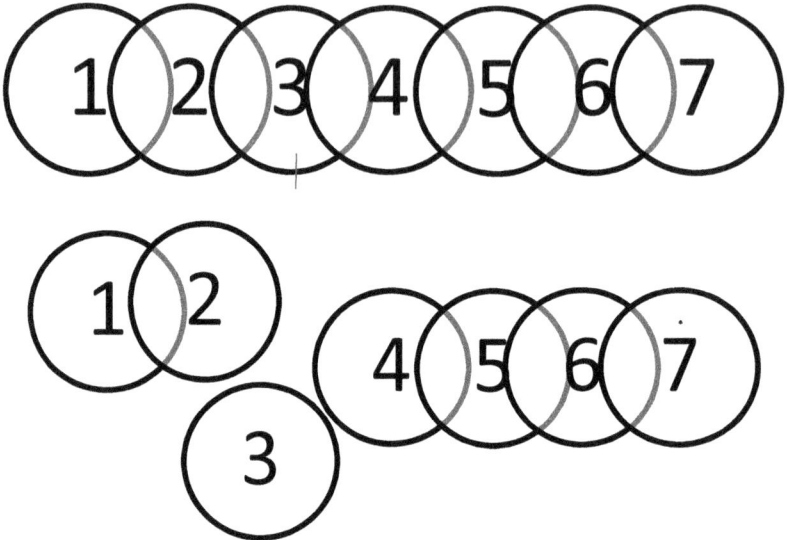

By applying common sense and some creativity, it's not hard to come upon the following payment schedule:

- Pay the third ring for the first night;
- Give the owner the first and the second connected rings on the second night and get the third ring back as change;
- Pay the third ring for the third night;
- Give the owner the fourth, the fifth, the sixth, and the seventh connected rings on the fourth night, and get the third ring and the first and the second connected rings back;
- Pay the third ring for the fifth night;
- Pay for the sixth night the same way as for the second night, give the first and the second connected rings and get the third ring back;
- Pay the third ring for the last night.

Simple logic was perhaps enough to come up with the sequence of payments illustrated above. But, as we mentioned before, there is an intriguing general method of explaining this solution that is applicable to more difficult cases, when the number of rings is greater than 7.

Binary representation of natural numbers provides an interesting and enlighting explanation of the above payment schedule.

Before going into the details, we would need to do some important explanations.

The base-2 numeral system is a positional notation with a radix of 2. Each digit is referred to as a bit and any number can be represented by a sequence of bits. In the binary system, each digit represents an increasing power of 2. The value of a binary number is the sum of the powers of 2 represented by each "1" digit.

We can express each of the numbers 1–7 as the following:

1:   01 because $1 = 1 \cdot 2^0$
2:   10 because $2 = 1 \cdot 2^1 + 0 \cdot 2^0$
3:   11 because $3 = 1 \cdot 2^1 + 1 \cdot 2^0$
4:  100 because $4 = 1 \cdot 2^2 + 0 \cdot 2^1 + 0 \cdot 2^0$
5:  101 because $5 = 1 \cdot 2^2 + 0 \cdot 2^1 + 1 \cdot 2^0$
6:  110 because $6 = 1 \cdot 2^2 + 1 \cdot 2^1 + 0 \cdot 2^0$
7:  111 because $7 = 1 \cdot 2^2 + 1 \cdot 2^1 + 1 \cdot 2^0$.

Applying the above expressions, we can now look at the payment schedule from the hotel owner's perspective:

- Day 1 — 01: He got one ring (right bit in our expression represents 1);
- Day 2 — 10: He got two rings and gave back one ring (left bit represents number 2 and the right bit 0 "stands for" no one-link rings);
- Day 3 — 11: He got one ring back (he has now two connected rings — left bit is 1, and 1 one-link ring — right bit is 1);
- Day 4 — 100: He got four connected rings and paid back two rings and one ring (the left bit is 1 representing four rings and the next two bits are 0 each — no two-linked rings and no one-link ring);
- Day 5 — 101: He got one ring (still has four rings in his possession — left bit is 1; and one-link ring — the last right bit is 1);
- Day 6 — 110: He got two rings and gave one ring back (the left 1 bit represents four rings, next 1 bit represents two rings, and the right 0 bit represents none of one-link rings);
- Day 7 — 111: He got one ring. We are done, all seven rings are paid to him (1 bit at each position represents, respectively, four rings, two rings, and one ring).

Many other difficult problems may be solved by applying a similar technique.

The base-2 numeral system is very useful in so many ways because every positive integer can be uniquely expressed as the sum of the powers of 2. This system is widely exploited in applied mathematics and it is used by almost all computer-based devices. The detailed discussion of this topic is beyond the scope of our book. The readers may wish to explore the implementations of binary numbers and the base-2 numeral system further. It should be challenging to investigate the generalizations of our problem, to consider different cases when the chain has more than seven rings and the hotel stay is longer than seven nights. For example, try to prove that one needs to make only two cuts to be able to pay one ring per night for a 17-day hotel stay and just three cuts for a 67-day stay. What about a general case scenario for any natural $n$? Can you suggest one?

The lesson learned from the problem is how important it is to properly interpret the problem's data and question asked. Sometimes, we make assumptions inconsistent with the known information. We do not properly apply the given attributes or even tend to misread and misunderstand what has to be used. This may result in solving basically a different problem. As soon as we clarify what is given and what has to be found or resolved, the rest of the problem's solution becomes managable. It was critical to grasp the meaning of the demand "to *pay* one ring per night" which helped to turn the whole thinking process in a proper direction.

In certain word problems, it looks like the given information directs you to the path of the solution and indicates what steps to take to get the result. However, the straightforward approach might get you nowhere; instead, you need to view a problem under a different angle, as it is advisable in the following Problems 5 and 6.

**Problem 5.** Two cyclists simultaneously started a training run facing each other from two cities located 300 miles apart. Starting on one cyclist's shoulder, a fly flew ahead to meet the other cyclist. Upon reaching him, the fly turned back and flew till it met the first guy. The fly shuttled back and forth between the two of them until they met, then she settled on the head of one of the cyclists. The fly's speed was 20 m/h. Each cyclist's speed was 15 m/h. How many miles did the fly travel?

**Solution.** The problem looks very confusing and even unsettling until one realizes that there is no need to count all the distances the fly covered shuttling back and forth between the cyclists. This is a typical motion problem. To find the distance, we need to know the speed (it is given to us) and the time spent to cover this distance with the given speed. Since the cyclists were facing each other (the critical detail in the problem!), and were riding with the same speed of 15 m/h, the distance covered by each till their meeting point was $300 \cdot \frac{1}{2} = 150$ miles. The time spent equals $150 : 15 = 10$ hours. But, this is the same time that the fly spent making her trips between them flying at the speed of 20 m/h. Therefore, the distance the fly traveled was $20 \cdot 10 = 200$ miles.

**Problem 6.** (Classic riddle about fair pay for one's share of the meal.) Two travelers met a stranger in a desert who told them that he was starving for several days, and he asked them to share a meal with him. The travelers planned to take a rest break, so they kindly shared the meal with the stranger. They ate five dishes (equal portions) in total. After finishing the lunch, the grateful stranger gave them five gold coins as his share of the price of the meal. How should the travelers fairly allocate the coins between them, if the first traveler bought three dishes and the second traveler bought two of the dishes that they ate together? Assume that each of the three men ate the same share of the meal.

**Solution.** The first natural desire is to suggest three gold coins to the first traveler and two coins to the second guy. However, this does not provide a fair distribution according to the problem's narrative. To solve this problem, one needs to understand that the stranger left *his share* of the total price paid for the meal. Since they shared five dishes equally, each was supposed to pay the same price of five coins for the meal. Hence, the total cost is $3 \cdot 5 = 15$ coins, and it follows that they paid 15 gold coins for five dishes. Hence, $15 : 5 = 3$ coins is the cost of one dish. We know that the first guy paid for three dishes, so he spent nine coins, and the second guy paid for two dishes spending six coins. Since each of them ate the same share of the meal, the fair distribution of the stranger's five gold coins has to be the following:

4 coins to the first traveler; he then would have paid in total
$$9 - 4 = 5 \text{ coins},$$

and

1 coin to the second traveler; his spending in total then equals
$$6 - 1 = 5 \text{ coins.}$$

These two cute problems give good examples of how erroneous one's first reaction can be, and how important it is to clearly understand what is known to us, and concentrate on crucial small details (sometimes, they seem irrelevant) before making any conclusions.

The following old classic problem belongs to what I call "perception" problems, in which the analysis of the given information is not enough to get the result. The surrounding circumstances or reactions of people involved create additional data that is instrumental to making decisions.

**Problem 7.** Three Masters of Logic wanted to find out who was the wisest among them. So, they turned to their Grand Master, asking him to resolve their dispute. "No problem", the old man said. "I will blindfold you and put a hat on each man's head. I have three white hats and one black hat. When I take your blindfolds off, the one who guesses the color of the hat on his had first, wins". When he took their blindfolds off, all three men sat in silence pondering. For a while none of them said a word, as nobody could figure out the color of the hats on their heads. Finally, one of them proudly announced that he has a white hat on his head. How did he think it through?

**Solution.** The critical detail in this problem is the fact that after opening their eyes, all the sages remained silent. The smartest one was the first to realize that if he would have a black hat on his head, the other two sages obviously would have seen it. Then, at least one of them should have been able to determine that there must be a white hat on his head because the only available black hat would be on the smartest sage's head, and he would immediately raise his voice. But, all three men were silently looking at each other. That clearly should mean that each man saw the same picture — a white hat on the other two heads. By analyzing the problem's data and the sages' behavior (which became the part of the problem!), the smartest one was the first to make the correct conclusion regarding the color of his hat.

Sometimes, making a diagram or a table representing a problem's data may be useful in clarifying what the problem is about; it may significantly simply the solution process.

We address it now, as you are likely to encounter many other instances in subsequent chapters.

**Problem 8.** (Suggested by a French mathematician Edouard Lucas (1882–1891).) Every day at noon, a ship leaves Le Havre for New York and another ship leaves New York for Le Havre. The trip lasts seven days and seven nights. How many New York — Le Havre ships will the ship leaving Le Havre today meet during its journey to New York?

**Solution.** The typical "obvious" response of seven ships is not correct.

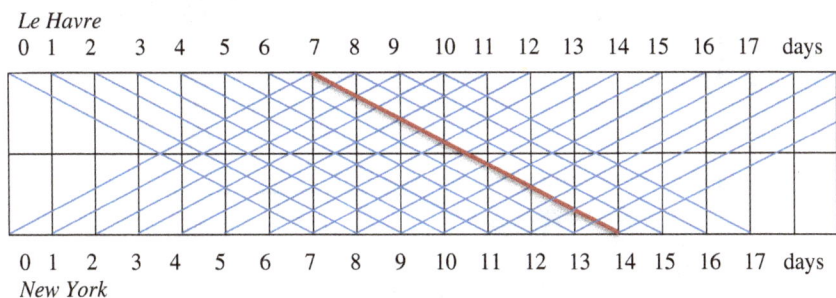

As our ship is leaving Le Havre today, on its seven days and seven nights' route to New York, it will indeed meet seven ships that have not left the harbor yet. But, when giving such an answer, one forgets about the ships that are already on route from New York to Le Havre. The vivid solution is displayed on the diagram above.

Let's consider a ship which is leaving Le Havre on the seventh day of a month. The ships that will arrive from New York to Le Havre on the same day left New York seven days ago. As our ship will be on its way from Le Havre to New York, every day it will meet a ship coming from New York to Le Havre, including the ships already on route. There will be a total of 15 meetings, counting 13 ships at sea and one in each harbor. The meetings are daily, at noon and midnight.

Referring to the diagrams, we will turn to "problems on sets", conveniently simplified by Venn diagrams called so after British mathematician John Venn (1834–1923).

Before we proceed to problem examples, we have to cover some introductory explanations.

A set is defined as a collection of distinct objects. Sets are usually denoted with capital letters. Sets $A$ and $B$ are equal if and only if they have precisely the same elements. The relationship among the members of sets can be represented by circles. The diagrams below depict various combinations of two sets. The interior of the circle represents the elements of the set, while the exterior represents elements that are not members of the set.

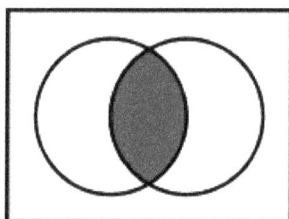

Intersection of two sets

(1)

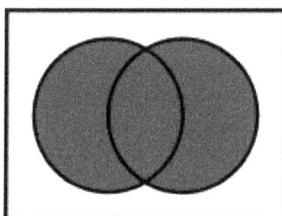

Union of two sets

(2)

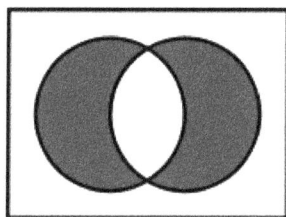

Symmetric difference of two sets

(3)

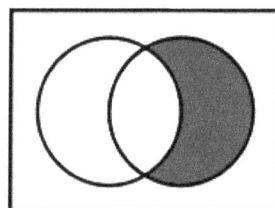

Relative complement of two sets

(4)

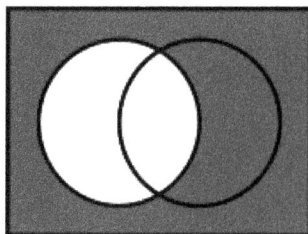

Absolute complement of one set in another set

(5)

A Venn diagram allows vivid visualizations. For example, an intersection of two sets, that is, the set of all elements that are common members of both sets $A$ and $B$, $A \cap B$ (see Figure 1), is represented visually by the area of overlap of the regions $A$ and $B$, while the union of two sets shown in Figure 2, $A \cup B$, represents the set consisting of all the elements of both sets. In Venn diagrams, the curves are overlapped in every possible way, showing all possible relations that exist between the sets, helping establish logical connections between sets' elements and drawing important conclusions.

**Problem 9.** In a group of 100 students, 52 students learn French and 70 students learn Spanish. How many learn French only? How many learn Spanish only? How many learn both languages?

**Solution.** The 100 students can be divided into three sets: those who study French only, those who study Spanish only, and those who study both languages.

Let $A$ be the set of students who learn French and $B$ be the set of students who learn Spanish.

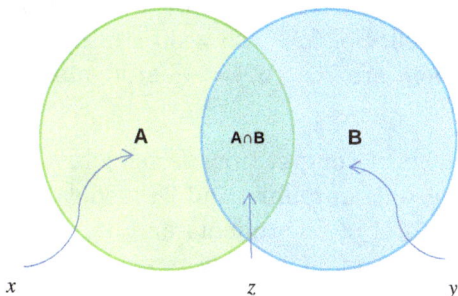

First, before going into the details, let's make several general observations. Assume that

$x$ — the number of items that belong to set $A$ only;
$y$ — the number of items that belong to set $B$ only;
$z$ — the number of items that belong to set $A$ and $B$ both (common elements of two sets).

From the Venn diagram, it is clear that $n(A) = x+z$, $n(B) = y+z$, $n(A \cap B) = z$. Therefore,

$$n(A \cup B) = x+y+z = (x+z)+(y+z)-z = n(A)+n(B)-n(A \cap B).$$

So, we have that

$$n(A \cup B) = n(A) + n(B) - n(A \cap B) \qquad (*)$$

The union of sets $A$ and $B$ consists of 100 students, $n(A \cup B) = 100$. To solve the problem, we will start with answering the last question first; we have to determine how many out of all 100 students learn both languages, i.e., what the intersection of two sets is, $n(A \cap B)$. This can be calculated from the above formula (*) as

$$n(A \cap B) = n(A) + n(B) - n(A \cup B) = 52 + 70 - 100 = 22.$$

The Venn diagram helps us visualize the problem's information. Both sets have a total of 100 students. However, by adding the number of students who learn French and number of students who learn Spanish gives $52 + 70 = 122$. Clearly, the excess of 122 over 100, 22 students, represents the number equal to double counting of students who learn both languages. This is depicted as the set of numbers $A \cap B$ in the diagram above. Finding the students who learn both languages allows us now to easily calculate the number of students who learn only one language. There are 52 students who learn French and 22 of them learn Spanish as well. So, $52 - 22 = 30$ students who learn French only. Likewise, there are $70 - 22 = 48$ students who learn Spanish only.

**Problem 10.** In sport competitions, a school is awarded 15 medals in volleyball, 10 medals in tennis, and 28 medals in football. If these medals went to a total of 40 students and only four individuals got medals in all three sports, how many received medals in exactly two of these categories?

**Solution.**

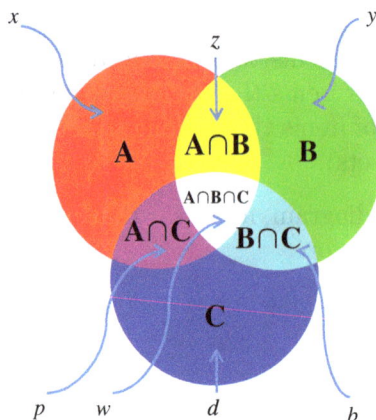

We will start with the similar analysis as in the previous problem. Assume that

$x$ — the number of items that belong to set $A$ only;
$y$ — the number of items that belong to set $B$ only;
$z$ — the number of items that belong to set $A$ and $B$ (common elements of two out of three sets only);
$d$ — the number of items that belong to set $C$ only;
$p$ — the number of items that belong to set $A$ and $C$ (common elements of two out of three sets only);
$b$ — the number of items that belong to set $C$ and $B$ (common elements of two out of three sets only);
$w$ — the number of items that belong to set $A$, $B$, and $C$ (common elements of three sets).

From the Venn diagram, it is clear that

$$n(A) = x + z + w + p,$$
$$n(B) = y + z + w + b,$$
$$n(C) = d + p + w + b,$$
$$n(A \cap B) = z + w, \ n(A \cap C) = p + w, \text{ and } n(B \cap C) = b + w,$$
$$n(A \cap B \cap C) = w.$$

We see from the diagram that $n(A \cup B \cup C) = x+z+w+p+y+b+d$.

We can modify the last equality as the following (we add and subtract the same numbers to get the specific sets):

$$
\begin{aligned}
n(A \cup B \cup C) &= (x+z+w+p) + (y+z+w+b) - z - w + d \\
&= (x+z+w+p) + (y+z+w+b) + (d+p+w+b) \\
&\quad -z-w-p-w-b \\
&= n(A) + n(B) + n(C) - (z+w) - (p+w) \\
&\quad -(b+w) + w \\
&= n(A) + n(B) + n(C) - n(A \cap B) - n(A \cap C) \\
&\quad -n(B \cap C) + n(A \cap B \cap C).
\end{aligned}
$$

So,

$$n(A \cup B \cup C) = n(A) + n(B) + n(C) - n(A \cap B) - n(A \cap C)$$
$$- n(B \cap C) + n(A \cap B \cap C).$$

Let $A$ be the set of students who got medals in volleyball, $B$ be the set of students who got medals in tennis, and $C$ be the set of students who got medals in football. It is given that $n(A) = 15$, $n(B) = 10$, $n(C) = 28$. The union of sets equals $n(A \cup B \cup C) = 40$, that is, the total of all students awarded with medals in the three sports. The intersection of sets, i.e., the number of students who got the medals in each of the three sports, equals $n(A \cap B \cap C) = 4$. Our goal is to find the number of elements belonging to exactly two of the three sets $A$, $B$, and $C$. As it is clearly evident on the Venn diagram, $A \cap B \cap C$ is the set of the common elements belonging to the sets $A \cap B$, $A \cap C$, and $B \cap C$. If we eliminate these common elements from each of the three sets, the sum of the remaining elements in each intersection will give us the total elements in the three sets, the desired outcome. So, the ultimate goal is to find

$$n(A \cap B) + n(A \cap C) + n(B \cap C) - 3n(A \cap B \cap C) \qquad (1)$$

Now, we can refer to the general equality we derived from the Venn diagram above:

$$n(A \cup B \cup C) = n(A) + n(B) + n(C) - n(A \cap B) - n(A \cap C)$$
$$- n(B \cap C) + n(A \cap B \cap C).$$

It follows that

$$n(A \cap B) + n(A \cap C) + n(B \cap C) = n(A) + n(B) + n(C)$$
$$+ n(A \cap B \cap C) - n(A \cup B \cup C).$$

Using the last equality and substituting the given values in (1), we obtain

$$n(A \cap B) + n(A \cap C) + n(B \cap C) - 3n(A \cap B \cap C)$$
$$= 15 + 10 + 28 + 4 - 40 - 12 = 5.$$

Five students received medals in exactly two of the three sports.

We will turn now to more complicated problems and will illustrate how the thorough and diligent analysis of the given information should lead to designing a problem-solving plan in the most efficient way.

**Problem 11.** One day, a young man and an older man left the suburb for the city, one on a bicycle, and one in a car. Soon, it became evident that if the older person had ridden three times as far as he had, he would have half as far to ride as he had, and if the young man had ridden half as far as he had, he would have three times as far to ride as he had. Who drove the car?

**Solution.** To answer the problem's question, we will need to compare the speed of every person who traveled. Clearly, the person who rode the car should be driving with a greater speed. However, there is nothing said about the speed at all. The only information given to us is the actual distances covered in comparison with some hypothetical scenarios. Bearing in mind that they left the suburb simultaneously, comparing the distances ridden by the travelers for the same time period would allow us to compare their speeds and make a conclusion about who was driving a car. To better understand the problem, we will use two diagrams. In doing so, it is important to keep in mind that both men covered the same distance. So, we consider two equal segments $AD$ and $A_1D_1$ representing, respectively, the older man's route and the younger man's route from the suburb to the city.

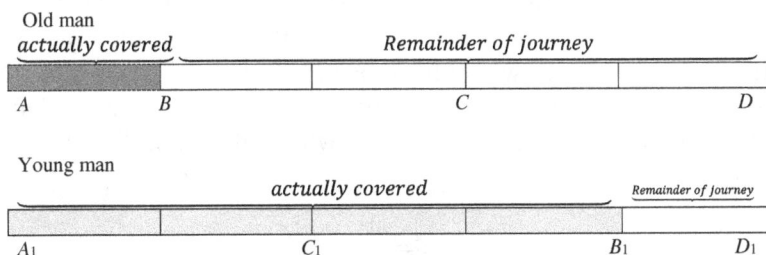

In the first diagram, the segment $AB$ represents the actual distance covered by the older man. We mark $C$ such that $AC = 3AB$. If the older man had ridden three times as far as he had, he would have ridden to $C$. Then, we mark $D$ such that $CD = \frac{1}{2}BD$, that is, the actual remainder of the journey is twice as big as the distance from $C$ to $D$. Note that in the above notations, we divided the full distance into five equal segments.

In the second diagram, $A_1B_1$ represents the distance actually covered by the young man. The midpoint $C_1$ of $A_1B_1$ indicates where the young man would be if he had ridden half as far as he had. Now, we mark $D_1$ such that $C_1D_1 = 3B_1D_1$, so the actual remaining distance is $\frac{1}{3}$ of the distance from $C_1$ to the end of the journey. Once again, the distance between the suburb and the city (the same for both men!) is divided into five equal parts. We see that the young person covered a distance four times what was covered by the older person. Since they started the journey at the same time, the young person was driving faster, and we arrive at the conclusion that he was the one who was driving the car.

Another way of solving the problem is to translate it into algebra terms and compare the distances actually covered by each person by expressing them through a common variable.

Let $x$ miles be the distance from the suburb to the city and $y$ miles be the distance actually covered by the old guy. Then, the remaining distance for him is $(x - y)$ miles. If he had ridden $3y$ miles, he would have had $(x - 3y)$ miles left. On the other hand, he would have left $\frac{1}{2}(x-y)$ miles. So, we get an equation $x - 3y = \frac{1}{2}(x-y)$. Simplifying, we obtain that

$$y = \frac{1}{5}x. \tag{2}$$

Let $z$ miles be the distance actually covered by the young person. The remaining distance then is $(x - z)$ miles. If he had ridden $\frac{1}{2}z$ miles, the remaining distance would have been $(x - \frac{1}{2}z)$ miles. On the other hand, it is known that he would then have had $3(x - z)$ miles left. It follows that $x - \frac{1}{2}z = 3(x - z)$, from which after simplifications we get that

$$z = \frac{4}{5}x. \tag{3}$$

Comparing expressions (2) and (3), we arrive at the same conclusion as from utilizing the diagrams before; the young person rode four times as far as the older person. He was driving with the higher speed, so he was the car driver.

The considered problem is regarded as difficult because of its confusing language. As soon as we clarified the given relationships, the problem became pretty straightforward and easy to solve. It is critical here to realize that the problem is not about calculating some specific values, but rather about comparing the distances covered. The word *compare* is the key to the solution. In the first solution, the diagrams enlightened the given particulars into vivid pictures and in the second solution, the equations helped to translate the data into algebra and find the relationships between the distances covered by both men.

If in the problem above, an algebraic solution was just one of the alternatives, then in many problems, it is the best and the most efficient way to simplify and interpret the conditions to find a path in the logical labyrinth.

**Problem 12.** A policeman arrived at a car accident scene and asked the witnesses for help. One of the vehicles that were involved in the accident had left and it was important to get information to identify the car. "I noticed that it was a two two-digit number on the plates. The second number can be obtained from the first by reversal of its digits and if you subtract the second number from the first, the result will be the sum of the digits of one of the numbers", said one of the witnesses of the accident. Help the policeman figure out the driver's plate numbers of the second vehicle.

**Solution.** To analyze the problem and utilize the info obtained from the witness, we need to introduce several variables and translate the details into algebra terms. Assume the two two-digit number on the plates was $ab - cd$. Each of the two-digit numbers can be written as $(10a + b)$ and $(10c + d)$ respectively. It is also known that $(10a + b) - (10c + d) = a + b$, from which we obtain that

$$10a + b - 10c - d = a + b,$$
$$9a - 10c - d = 0.$$

The witness said that $a = d$ and $b = c$. Substituting $a = d$ into the last equality gives $8a = 10c$, or $a = \frac{5}{4}c$.

Since $a$ and $c$ each represents a single-digit number, the only values $c$ can get are 4 and 8. If $c = 8$, then $a = \frac{5}{4} \cdot 8 = 10$, which is impossible because $a$ has to be a single-digit number. So, the only available

option remaining for $c$ is $c = 4$. Then, $a = 5$ and $b = c = 4$, $d = a = 5$.
We arrive at the desired number on the plates to be "54 – 45".

**Problem 13.** Given a number $19*83$, substitute a digit for a $*$ in
the middle of the number to get the perfect cube.

**Solution.** This problem looks tough, so the advice is to work it
one step at a time and start with what we know. This is a good
example of building a logical chain determining every next step based
on previous decisions. First of all, since our goal is to get the perfect
cube, let's introduce $x$ such that $19*83 = x^3$.

Second, note that we are dealing with a five-digit number.

Obviously, $x$ cannot be a one-digit number because
$10^3 = 1,000 < 19*83$; $x$ cannot be a three-digit number
either because $100^3 = 1,000,000 > 19*83$. Therefore, it has to
be a two-digit number and such that $20 < x < 30$ because
$8000 < x^3 < 27,000$. Now, it is time to pay more close attention
to the given number $19*83$ and the digits it consists of, specifically
its last digit 3. Out of all nine digits, the only cube of 7 has 3 as its
last digit. Recalling the determined restrictions on $x$, $20 < x < 30$, we
arrive at the only possible outcome of $x = 27$. Indeed, $27^3 = 19,683$.
Therefore, the desired digit to substitute for a $*$ is 6.

It's time now to introduce several problems, for which the hint to
a solution is already embedded in the problems' narrative.

*There is something about Mary.*
Mary's mom has four children.
The first child's name is April.
The second is May.
The third is June.
What is the name of the fourth child?

Did you immediately realize the fourth child's name? If, yes, then
good for you! Many times, however, after reading this problem to my
students, I've seen many of them became confused by the context of
the problem. After reading it a second or even a third time though,
the students' laughing answer was "Of course, it's Mary!"

It takes just a quick mental check to see that the answer to the
above riddle is embedded in its language. In the next few problems,
this is not as evident as in this example, but they all have one thing

in common; the way the problems are presented to us provides a major clue for simplifying the solution.

**Problem 14.** Simplify $\frac{(x-a)(x-b)}{(c-a)(c-b)} + \frac{(x-a)(x-c)}{(b-a)(b-c)} + \frac{(x-b)(x-c)}{(a-b)(a-c)}$, where $a \neq b$, $b \neq c$, $a \neq c$.

**Solution.** As many problems considered in the book, this one has several solutions. One of the next chapters will be fully devoted to reviewing different solutions, comparisons, and conclusions from the alternative methods and techniques applied to various selected problems. This problem is a good example of when a straightforward approach to a solution appears to be not the best option. The tedious, but perhaps the most apparent, way to simplify this expression is in finding the common denominator with further manipulations of the expression obtained in the nominator. We invite readers to try it on their own and then compare the effectiveness with the much simpler approach suggested below. In this book, we are dealing mostly with non-standard problems, so general problem-solving strategies do not always provide the best results. There are many artificial tricks and techniques allowing for more elegant and efficient solutions.

There are multiple reasons for adding auxiliary elements as supplemental tools in many problems. Sometimes, we use them to link some elements together and utilize new relationships not seen before; sometimes, we need them to clarify the picture and make it more vivid and explicit. Sometimes, auxiliary elements are helpful in establishing an analogy with previously solved problems. Sometimes, they just significantly simplify the solution process and are critical in finding an elegant and beautiful solution. In our case here, the given expression to simplify is an open invitation to introduce an auxiliary function $f(x) = \frac{(x-a)(x-b)}{(c-a)(c-b)} + \frac{(x-a)(x-c)}{(b-a)(b-c)} + \frac{(x-b)(x-c)}{(a-b)(a-c)}$. From the way it is defined, it is clear that $f(x)$ is either a linear or a quadratic function (the highest power of the variable after all simplifications can't be greater than 2).

Let's find its value for the three particular values of $x$ when $x = a$, $x = b$, and $x = c$:

$$f(a) = 0 + 0 + 1 = 1,$$
$$f(b) = 0 + 1 + 0 = 1,$$
$$f(c) = 1 + 0 + 0 = 1.$$

We see that $f(a) = f(b) = f(c) = 1$. This is possible only when $f(x) = 1$ for any value of $x$. It implies that $\frac{(x-a)(x-b)}{(c-a)(c-b)} + \frac{(x-a)(x-c)}{(b-a)(b-c)} + \frac{(x-b)(x-c)}{(a-b)(a-c)} = 1$.

The problem is solved without any tedious calculations and modifications!

**Problem 15.** Evaluate without a calculator
$$\frac{1234567890}{1234567891^2 - 1234567890 \cdot 1234567892}.$$

**Solution.** Working with many so-called "integer" problems involving evaluations of expressions with big integers or word problems involving determinations of some integer amounts based on specific conditions, it is worthwhile to introduce an auxiliary variable. An appropriate selection for such a variable usually significantly simplifies the entire picture and allows us to contemplate a plan for a solution. The hint on how to introduce the auxiliary element in our case is hidden in the conditions.

Indeed, letting $x$ be $x = 1234567891$, we obtain that $x - 1 = 1234567890$ and $x + 1 = 1234567892$. The given fraction then simplifies to

$$\frac{1234567890}{1234567891^2 - 1234567890 \cdot 1234567892} = \frac{x-1}{x^2 - (x-1)(x+1)}$$

$$= \frac{x-1}{x^2 - (x^2 - 1)} = \frac{x-1}{x^2 - x^2 + 1} = x - 1 = 1234567890.$$

Usually facing similar problems, students hesitate to convert to algebra terms and make attempts to solving a problem directly working with numbers. Sometimes, it does work; sometimes, it becomes a tedious, frustrating, and worthless process. Do you need to get some creativity to come up with an idea of introducing a variable as we did it? Yes, absolutely. But, in this case, it was not that hard as soon as you start thinking in the direction of substituting numbers for variables. The big numbers given in the nominator and denominator differ just by 1, so why not try to simplify it working with variables instead of numbers. In many similar problems, this idea may not immediately lead to the desired outcome, but at least it may clarify the path to a solution and simplify the whole process.

**Problem 16.** Solve the equation $(x^2 - 4x + 5)(z^2 + 8z + 20) = 4$.

**Solution.** This is a multi-variable equation. Any conventional approaches for single-variable equations' solutions are bound to fail here. While solving many equations, we are often expanding and then simplifying the given expressions. It is not the case here. Expanding the expression on the left-hand side will make the equation even more confusing. We see that each factor on the left-hand side is a quadratic trinomial. What about if we recall the properties of a quadratic function and try to apply them to analyze the problem?

According to the properties of a quadratic function $f(x) = ax^2 + bx + c$, for $a \neq 0$, the minimum value of each factor will be attained at the vertex of the respective parabola for each quadratic function (each parabola opens upward because the first coefficient is a positive number).

The graphical illustration of the general case is the following:

$a > 0$
$D > 0$ ($D$-discriminant)
Two points of intersection.
Two roots.

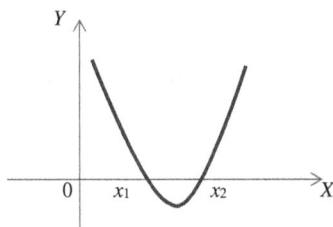

Let's consider the function $f(x) = x^2 - 4x + 5$. The abscissa of the parabola's vertex is $x = \frac{-b}{2a} = \frac{4}{2} = 2$. The function has a minimum that equals $f(2) = 2^2 - 4 \cdot 2 + 5 = 1$. For the second function $f(z) = z^2 + 8z + 20$, its minimum value is attained when $z = \frac{-b}{2a} = \frac{-8}{2} = -4$ and equals $f(-4) = (-4)^2 + 8 \cdot (-4) + 20 = 4$. Therefore, the left-hand side of the given equation exceeds or equals $1 \cdot 4 = 4$. The equality is attained when $x = 2$ and $z = -4$. Thus, the only possible solutions of the given equation are $x = 2$, $z = -4$.

As we see, in this case, the problem's specifics, or in other words, the way the equation is written, provide a hint on how we can approach the solution. What else could you logically do, other than to exploit the quadratic function's properties here? It paved the way for a pretty simple solution. The advice "do whatever is doable" happens to be very useful in this case.

**Answer:** $x = 2$, $z = -4$.

**Problem 17.** Find $x$ and $y$ if $\frac{x-3}{xy-2} = \frac{y-5}{xy-3} = \frac{8-x-y}{5-x^2-y^2}$.

**Solution.** The obvious first desire is to rewrite the above expression as a system of two equations with two variables; it suffices to consider any two equalities out of the given three to get such a system. Solving the system, we will find $x$ and $y$. It is a valid suggestion but not the best one in this case. The way the problem is presented to us provides a hint for an easier and more elegant way to get the desired results. Having three equal ratios, we can introduce another variable $t$ as equal to each fraction. It follows that

$$\begin{cases} x - 3 = t(xy - 2), \\ y - 5 = t(xy - 3), \\ 8 - x - y = t(5 - x^2 - y^2). \end{cases}$$

Now, adding three equations in the system, we obtain that

$$x - 3 + y - 5 + 8 - x - y = t(xy - 2 + xy - 3 + 5 - x^2 - y^2)$$

After simplifications, we get the equation $t(2xy - x^2 - y^2) = 0$ or equivalently,

$$t(x - y)^2 = 0.$$

Therefore, either $t = 0$ or $x = y$.

If $t = 0$, substituting it in the first and the second equation, we get, respectively,

$$x = 3, \ y = 5.$$

In case $x = y$, substituting $y$ for $x$ into the first equation $\frac{x-3}{xy-2} = \frac{y-5}{xy-3}$ gives $\frac{x-3}{x^2-2} = \frac{x-5}{x^2-3}$, from which $(x-3)(x^2-3) = (x-5)(x^2-2)$. This simplifies to $x^3 - 3x^2 - 3x + 9 = x^3 - 5x^2 - 2x + 10$, and finally, to $2x^2 - x - 1 = 0$. This quadratic equation can be solved by applying the general formula for its roots.

$$D = (-1)^2 + 4 \cdot 2 \cdot 1 = 9,$$

$$x = \frac{1 \pm \sqrt{9}}{4}.$$

We get that $x_1 = 1$, $x_2 = -\frac{1}{2}$. Therefore, $y_1 = 1$, $y_2 = -\frac{1}{2}$.

It is easy to verify that all found values for $x$ and $y$ satisfy the conditions $xy \neq 2$, $xy \neq 3$, and $5 - x^2 - y^2 \neq 0$ (the restrictions on the domains of each expression in the system). So, we found three pairs of the solutions $(3, 5)$, $(1, 1)$, $(-\frac{1}{2}, -\frac{1}{2})$.

**Answer:** $(3, 5)$, $(1, 1)$, $(-\frac{1}{2}, -\frac{1}{2})$.

The best detective stories and novels written by prominent authors such as Arthur Conan Doyle, Agatha Christie, Wilkie Collins, and Georges Simenon (readers can extend the list with names of their favorite authors) survived the test of time and are as popular today as in the past. The main reason for this is not just because readers are captivated by the plot intrigues. While reading the stories, we try to solve the mysteries on our own, analyze the given information, and build logical chains leading to our own conclusions about who the criminals are. The whole process, as a matter of fact, is exactly the same as the problem-solving process.

Let's conclude this chapter devoted to the analysis of a problem data with a detective story. It provides a great example of how important the ability to thoroughly analyze the given conditions of a problem can be.

**Problem 18.**

*A Detective Story*

A small suburban town was shocked with terrible news about an early morning cold-blooded robbery of a local jewelry store. The store owner was wounded with several gun shots and was unconscious; the robber managed to escape. Police inspector Scholms arrived at the crime scene and started questioning several witnesses who agreed to help him.

"I saw a tall muscular young guy wearing jeans and a blue t-shirt who was running away from the store", said Mr. Lemon.

"No, there was no one running from the store with such a description. I am positive that the robber was a skinny, bold, very short, middle-age guy in a black jacket. He jumped out from the window on the left side of the store", announced Mr. Jones.

"I can assure you, inspector, that both of these statements are not true. None of these two guys said a word of truth. I saw the robber when he stepped out from the door. The guy was of normal height, he had blond hair, and he was wearing a green jacket", the last witness Mr. Rich said.

"Did you see anybody, Mrs. Tarple?" the inspector asked an old lady who was standing by listening to the three witnesses.

"No, Sir. I just came here. I did not see anything", she said.

"Inspector, I have to say that Mr. Jones lied. Everything he said is just complete non-sense", said Mr. Lemon.

"Whatever Mr. Rich was saying about this blonde guy in a green jacket is not true at all!" said Mr. Jones.

"That's enough! All of you!" inspector Scholms yelled at the witnesses. He looked frustrated and puzzled.

"Inspector, I think that the information given to you is not completely worthless. One of them perhaps was telling the truth and he appears to be the most reliable of three". Mrs. Tarple said in a quiet and calm voice.

Which guy was the most trustworthy? How did Mrs. Tarple identify him?

**Solution.** Let's put together and analyze the information we have.

Jones said that Rich is a liar,

Lemon said that Jones is a liar,

Rich assured that both, Jones and Lemon lied.

First, assume that all three witnesses lied. Clearly, this is impossible, because Rich testified that Jones and Lemon lied. If Rich lied, then at least either Jones or Lemon was telling the truth.

Assume now that Rich did not lie. He stated that Lemon and Jones lied, and this has to be a true statement in our assumption. But, we also know that Lemon assured that Jones is a liar, which means Lemon lied that Jones is a liar. It implies that we get a contradiction (otherwise, Jones had to be the one who was telling the truth and the assumption that Rich did not lie is false). So, clearly, Rich is a liar.

Now, let's see what Lemon's claim was. Lemon assured that Jones is a liar. If this is a true statement, then it will support Rich's similar claim about Jones. But, we just established that this is not true. Therefore, Lemon's claim is false. So, the only guy who was telling the truth in his claim about the other two guys' claims (i.e., Rich being a liar), is Jones.

Another way to identify the person telling the truth about two other guys is the following. If we assume that Rich lied and one of the other two guys was telling the truth, we get that it has to be Jones who was telling the truth (he said that Rich is a liar), while Lemon lied about Jones being a liar.

So, in either case, we figured out that only Jones was telling the truth about the statements of the other two witnesses, while the other two guys lied.

What is interesting about this problem is that we do not know that each person either always told the truth or always lied. But, as Mrs. Tarple said, the inspector now realized who the most reliable person is, and he should concentrate on verifying the accuracy of Jones's statement about the criminal.

A keen use of logic and proper analyses of witnesses' testimonies allowed Mrs. Tarple to help the inspector get the important information about the suspect.

**Practice exercises.**

**Problem 19.** Prove that the number $1\underbrace{00\cdots00}_{49\ \text{zeros}}2\underbrace{00\cdots00}_{99\ \text{zeros}}1$ is not a cube of any integer.

**Problem 20.** A customer purchased goods for $10 and gave a $50 bill to the salesperson. The salesperson did not have change and asked his associate to exchange the $50 bill for him. After the customer left, the sales associate realized that the $50 bill he got from the salesperson was a fake bill and asked to reimburse him for it. The salesperson returned the $50 borrowed from his associate and started thinking about the loss he needs to report. How much of the loss did he incur?

**Problem 21.** Solve the equation $\frac{x-49}{50} + \frac{x-50}{49} = \frac{49}{x-50} + \frac{50}{x-49}$.

**Problem 22.** The sum of ages of all family members, consisting of mother, father, and son is 65 years. Four years ago, the father was nine times older than his son. Nine years ago, the sum of the ages of all family members was 40. How old is the father?

**Problem 23.** While getting his car fixed, for a few weeks, a father would come home from work by bus at 8 pm. Each day at 8 pm, his son would wait for him at the parking lot next to the bus station and would drive him home. One day, the father returned earlier at 7.30 pm and decided to walk home from the parking lot. Shortly after, the son left his house on the normal schedule, and picked his father up somewhere on the path from the bus station. They arrived back home 10 minutes earlier than usual. At what time did the son meet and pick up his father on his way from the parking lot?

**Problem 24.** A student wrote a multiplication problem of two two-digit prime numbers on the blackboard. After finishing the calculations, he substituted the digits of all numbers with letters (same digit is expressed by the same letter) and wrote the following equality:

$$\overline{AB} \cdot \overline{DC} = \overline{KUKU}.$$

Prove that he made an error.

**Problem 25.** Solve the equation

$$(x^2 + 3x - 4)^3 + (2x^2 - 5x + 3)^3 = (3x^2 - 2x - 1)^3.$$

# Chapter 3

# The Clue is in the Question

*Think on the end before you begin.*

— George Polya

When solving a problem, it is very important to clearly understand not just the given information but what the question is asking. If you do not understand what to determine or find, you could be solving an absolutely different problem.

It may even take funny anecdotic forms. For instance, the assignment given to a student was

*To expand the expression* $(a + b)^3$.

Instead of $(a + b)^3 = a^3 + 3a^2b + 3ab^2 + b^3$, the answer was given as the following:

$$(a + b)^3$$

$$(a \quad + \quad b)^3$$

$$(a \quad\quad + \quad\quad b)^3.$$

The selection of problems in Chapters 2 and 3 is debatable and to some extent even equivocal. In any problem, it is critical to understand its conditions and the question asked, so you can't really separate those elements. For example, the problem about the gold chain cut considered in the previous chapter might very well be suited to this chapter as well. It was critical to comprehend the difference between paying one ring for a hotel day stay and physically giving one ring per day to the hotel owner. However, this did not point out the plan for the solution. In this chapter, we will concentrate on problems that have questions that direct the path to their solutions. Very often, the hint or even a major key to a solution is hidden in the question asked. It may provide good directions pointing out a route for a problem's solution. It is also critical in solving a problem to always answer the question that is being asked, not something else, which can easily be confused with a desired outcome. Sometimes, you do not even need to "solve" a problem, just correctly understand what is asked and give the adequate response.

Let's go over several examples of seemingly confusing questions (dare we consider these examples as problems):

**Question 1.** Some months have 30 days and some have 31 days. How many months have 28 days?

Was the first thought that popped in your head that it could only be "just one month — February"? However, this is the wrong answer. Read the question again; you have to give a response to what is asked, not what you believe is being asked. The correct answer is — there are 28 days in *any month* of a year.

**Question 2.** Two friends played a chess game for 2 hours. How much time did each friend spend for this game?

Clearly, no calculations are needed for this riddle; each person played for 2 hours.

**Question 3.** One car was driving at the speed of 55 m/h going from city $A$ to city $B$. Another car was driving at the speed of 60 m/h going from city $B$ to city $A$. Which car will be further from city $A$ when they meet?

This question is a little more confusing than the previous two. But, it also does not require any mathematics to be involved. Obviously, when they meet, both cars will be at the *same distance* from city $A$.

**Question 4.** If it is rainy at 12 midnight, can you expect shiny weather in 72 hours?

Here, we finally have to do some math, $72 = 3 \cdot 24$. Three full days will pass. So, how can we expect shiny weather if it's going to be 12 midnight again? The question is about the sun at a specific time of the day. Clearly, you can't expect to have sun at midnight.

**Question 5.** (Back in the Soviet Union times, there were big lines in supermarkets to buy almost any type of products, so this one is an old real question faced by store visitors.)

One has to stand in a 20-meters line to buy a bologna sausage. The line to buy the butter is 5 meters longer than the first line, and the line to buy the bread is twice shorter than the first line. How many meters does one have to cover to be able to make the sandwich with the bologna sausage?

Pay attention to the question asked! It is about sandwich with sausage. Butter was not mentioned. So, the answer is $20 + \frac{20}{2} = 30$ meters.

I think you should be persuaded now to be very specific with answering a question asked and be more cautious before doing it. So, let's turn to examining problems, and start with the following two simple instances which also provide a good and vivid illustration of the concept.

**Problem 1.** A small land slug (shell-less snail) climbing a 10-ft tree makes it up the tree 3ft a day. But, being very tired, at the end of the day she slips back 2ft. At what day would she reach the top of the tree?

**Solution.** An obvious response, "on the 10th day", is wrong, because it is an answer to a question in how many days this slug would be able to cover 10ft, not exactly the question asked in the problem. The correct answer is, on the eighth day. Indeed, with the way she moves, she is going up 1ft/day. So, she will be at the 7-ft height at the end of the seventh day. Climbing up during the eighth day, she will reach the top of the tree by the end of the day just before she drops back down 2ft. Since the question was "At what day would she reach the top of the tree?", the correct answer is on the eighth day.

**Problem 2.** The supporting crew responsible for preparing the equipment for the NBA three-point contest did not properly count the number of basketballs needed. With a short amount of the time left before the event starts, they have to put aside 50 additional balls from the available 70 balls in the storage room. Assuming it takes 1 second to put aside one ball, how quickly would they be able to put aside 50 balls out of 70 available?

**Solution.** Don't rush to give an immediate response — 50 seconds. How about 20 seconds? The goal is to get the specific number of balls as quickly as possible. By counting 20 balls out of 70 available, we get the remaining 50 balls. So, we get the desired number of balls in 20 seconds!

The next problem belongs to the so-called clock-related problems. These problems are often regarded as difficult by many students because they do not properly interpret the question asked.

**Problem 3.** What is the angle between the hands of a clock at 6.34 o'clock?

**Solution.** Even though the question is about the angle between the hands of a clock at a specific time, it is important to understand that the goal is to find the difference of the distances covered by each hand of a clock for the indicated time period.

As the hands move, the angle they form will change. The problem is about the distance between them (in minutes) after specific time passed. In this type of "motion problem", the distance each hand travels is measured by the minute markers of the clock.

The hour hand of a clock turns at a rate of $360° : 12$ h $= 30°$/h or $0.5°$/min. The minute hand turns at a rate of $360° : 60$ min $= 6°$/min. Therefore, in 34 minutes, the minute hand will cover a distance of $34 \cdot 6° = 204°$; that is, the angle between the minute hand and the direction of 6 o'clock is $204° - 180° = 24°$. The hour hand will cover a distance of $34 \cdot 0.5° = 17°$ moving in a direction from 6 o'clock to 7 o'clock. In other words, the angle between the hour hand and the direction of 6 o'clock is $17°$. Hence, the angle between the hour and minute hands will be $24° - 17° = 7°$ at 6.34 o'clock. Clearly understanding and properly interpreting the question asked in terms of distances covered by each hand depending on their speed differences led the way for designing and executing the plan for the problem's solution.

I strongly suggest trying the more complicated problems offered below on your own before reading their solutions. The hint would be to pay extra attention to the question asked.

**Problem 4.** Two friends are driving from their houses facing each other. One is traveling at the speed of 60 m/h; another is traveling at the speed of 40 m/h. How far apart are they 1 hour before they meet?

**Solution.** There is not much given in the problem and it may look baffling how one can relate the speeds of the friends to the distance between them 1 hour prior to their meeting. The critical detail here is the fact that they are driving *facing each other*.

Solving this problem is important to concentrate on the question that is being asked. We don't care how long they will travel or how much of a distance will be covered by each 1 hour prior to their meeting. At some point in time, they will be apart 100 miles because they are facing each other. So, what will happen with each of them as they pass that point?

They will continue driving at the same speeds as before. We can rephrase the question asked to this: *what distance will each friend cover in 1 hour?* This clarifies the next steps that we have to take right away.

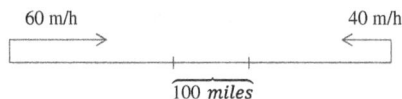

In 1 hour, the first guy will cover the distance of
$60\frac{m}{h} \cdot 1h = 60$ miles and the second guy will cover $40\frac{m}{h} \cdot 1\,h = 40$ miles. Since they are driving facing each other, the distance between them will be $60 + 40 = 100$ miles 1 hour prior to their meeting.

**Answer:** 100 miles.

You need to clearly understand what is demanded from you in the specific problem and answer only the question being asked. If you can find a shortcut and proceed directly to the requested item or event, good for you. Just go for it!

**Problem 5.** During the joint meeting between US and Russian astronauts at a space research scientific conference, the members of both delegations noticed that many of them met several times in the past. As it turned out, each member of the US delegation was acquainted previously with five members of the Russian delegation, and each Russian astronaut was acquainted previously with seven members of the American delegation. Which delegation has more people, American or Russian?

**Solution.** This cute problem looks very confusing at the first glance. From the given information, it's not clear how to figure out the number of members in each delegation. But, as a matter of fact, you don't need to know those numbers to answer the problem's question. We need to *compare* the numbers, which does not necessarily mean to find them.

Denote by $X$ the number of members of the US delegation and by $Y$ the number of members of the Russian delegation. Let's assume now that the astronauts from the two different countries will exchange small presents with their previously met counterparts (that's what usually happens on such meetings in real life). We see that there will be $5X$ gifts given out by US astronauts. On the other hand, since each Russian astronaut gets a gift from seven Americans, there will be $7Y$ gifts received. Since the gifts given and received represent the same quantity, we get an equation

$$5X = 7Y.$$

Obviously, it will hold only when $X > Y$. The problem is solved; there must be more people in the American delegation than in the Russian delegation.

This is an example of a problem, in which one benefits by analyzing the problem's data from the very end — from the question. As soon as you grasp that we need to *compare* two quantities, it may significantly simplify the solution, and provide a hint for what steps we next have to take. It is a different task to compare some numbers based on the known information rather than to calculate them, which, by the way, is impossible in this case. It merits also emphasizing the trick with introducing an auxiliary element, the gifts exchanged by the astronauts, which allowed getting a short and elegant solution.

**Answer:** The American delegation has more people.

After getting acquainted with the ideas applied in the solution of Problem 5, it should be easier for the readers to approach the following problem, which was offered by A. Spivak in the currently defunct magazine *Quantum*, September/October 1997 issue.

**Problem 6.** Every seventh mathematician is a philosopher, and every ninth philosopher is a mathematician. So, which is more numerous, mathematicians or philosophers?

**Solution.** Once again, we have to compare two numbers. Clearly, the key to the problem is to express through variables both sets, mathematicians and philosophers, and then analyze the findings. Contrary to the previous problem, we have common elements in both sets; there are people who belong to both sets, namely, mathematicians who are philosophers at the same time. Let's denote by $x$ the number of people who are mathematicians and philosophers. Hence, the number of mathematicians will be $7x$ and the number of philosophers will be $9x$. Indeed, each seventh of $7x$ is $x$, the number of mathematicians who are philosophers, and similarly, each ninth of $9x$ is $x$ as well, the philosophers who are mathematicians, i.e., it is the same number $x$. Since $9x > 7x$, we arrive at the conclusion that philosophers are more numerous.

In some problems, the question asked serves as a hint for where to start your pathfinding to an answer; it points you in the direction to design a plan for a solution.

**Problem 7.** Two brothers, Alex and Bryan, exchanged money between each of them. First, Alex gave Bryan some amount of his funds. Then, Bryan gave money to Alex. After that, Alex again gave money to Bryan. Finally, Bryan returned some amount back to Alex. After the final exchange, each brother ended up with $160. The amount given up in each exchange was equal to the amount of money on hand of the receiving brother prior to the exchange. How much money did each brother originally have?

**Solution.** Don't be fooled by the confusing language of this problem. Before starting any attempts answering the question, clarify the details and make a plan for how to approach the problem. We do not know how much money was exchanged every time. But, it is given that at the end the total amount they had was $320. Since no new funds have been involved, we can conclude that $320 is the constant cumulative amount on hands after each transaction. We also know that every time the amount given to the receiving person was exactly the same as the amount that person had, which translates into the fact that every time the money was doubled in the hands of the receiving person. This is a very important aspect helping us to analyze all the following steps in the solution process. Bearing in mind that the goal is to figure out the amount of money each brother had originally, the direction in which we have to start appears obvious, start the analysis from the very last exchange. Indeed, after the final exchange, Alex had $160. It implies that he was given $80, the exact amount of $80 he possessed just before getting the money from Bryan. Therefore, Bryan had to have $320 − $80 = $240. By the same logic, we can now rewind all the money transfers made between the brothers:

after the third exchange, Bryan's money doubled to $240, so he had $120 and Alex had $320 − $120 = $200;
after the second exchange, Alex's money doubled, so he had $100 and Bryan had $320 − $100 = $220;
after the first exchange, Bryan's funds doubled, so he had originally $110, and Alex then had originally $320 − $110 = $210.

Our findings can be organized in a table that provides a vivid and clear picture of all the transactions:

| | Money in possession | |
| Transfers | Alex | Bryan |
| --- | --- | --- |
| Original possession | $210 | $110 |
| 1st transfer $110 | $100 | $220 |
| 2nd transfer $100 | $200 | $120 |
| 3rd transfer $120 | $80 | $240 |
| 4th transfer $80 | $160 | $160 |

By interpreting the problem's particulars and concentrating on the question asked, we come up with finding a proper path for this logical labyrinth by going in the opposite direction, from the last transaction to the opening one.

**Problem 8.** Four girls, Veronica, Elly, Jessica, and Liana, participated in a school concert singing songs. Each song was performed by three girls. Veronica performed in eight songs, more than anybody else, and Elly performed in five songs, less than anybody else. How many songs did the girls sing?

**Solution.** It's one of those tricky problems, which is usually regarded as difficult and even unsolvable. It gives an impression that not enough information is given to us. However, as soon as we focus on the question asked, it should help to shed some light on how to approach the problem and how to design a plan for its solution. Indeed, all we need to find is the total number of songs performed by the girls no matter in what order and in what combinations it was done by any three of the girls among the four participants. Let's denote by $x$ the total songs they sang. We know that each song was performed by three girls. Somehow, we have to utilize this fact. To avoid tedious considerations of all possible combinations of three out of four performances (this may be not even useful at all), we can simplify the task by introducing an auxiliary element as a prize given to each girl for each performing song. So, assume that after each song is completed, every girl who was singing the song was awarded a prize. It follows that there should be a total of $3x$ prizes given. Now, we have to review the possible number of songs performed by Jessica and Liana. Since the greatest and the least number of songs performed by one person are 8 and 5, respectively, Jessica and Liana each could have performed in either 6 or 7 songs. Then, the total number of awards received should be one of the following:

$$8 + 5 + 6 + 7 = 26, \text{ or}$$
$$8 + 5 + 6 + 6 = 25, \text{ or}$$
$$8 + 5 + 7 + 7 = 27.$$

Since the total number of prizes awarded and distributed should be $3x$, some number divisible by 3, then the only possible outcome is the last one, 27 awards. So, we get a very simple equation, $3x = 27$, from which $x = 9$, i.e., the girls sang a total of 9 songs.

**Answer:** 9 songs.

**Problem 9.** Prove that for any natural $n$ the number $P = n^2 + n + 1$ is an odd number and there is no natural number such that its square equals $P$.

**Solution.** First, we need to prove that $P$ is an odd number no matter what values a natural $n$ takes. Let's slightly modify $P$ as $P = n^2 + n + 1 = n(n+1) + 1$ and analyze the last expression.

Assume $n$ is an even number, i.e., $n = 2m$, $m \epsilon N$, then $n(n+1) + 1 = 2m(2m+1) + 1$ is an odd number as the sum of even number $2m(2m+1)$ (it is even because it is divisible by 2, as one of its factors is 2) and 1.

Assume now that $n$ is an odd number, i.e., $n = 2m+1$, $m \epsilon N$, then $n(n+1)+1 = (2m+1)(2m+2)+1 = 2(2m+1)(m+1)+1$ is an odd number again, because it is the sum of even number $2(2m+1)(m+1)$ and 1.

To prove the second part of the problem, we need to work backwards. In other words, the question that is asked (the statement that has to be proved) is pointing us in the direction for our plan design.

Clearly, since $n$ is a natural number, $n^2 + n + 1 > n^2$. On the other hand, $n^2 + n + 1 < n^2 + 2n + 1 = (n+1)^2$. Combining these two facts, we can state that

$$n^2 < n^2 + n + 1 < (n+1)^2.$$

So, we see that the closest natural squares to our $P$ are the squares of two consecutive natural numbers, which implies that it is impossible for $P$ to be a square of some natural number for any $n \epsilon N$.

It worth emphasizing here that we assessed the lower and upper bound for $P$ in terms of the squares of natural numbers. I would say that this idea to compare $P$ with the closest natural squares came as an intrinsic consequence of the fact to be proved. It directed our way of thinking, allowing for a short and elegant arrival at the desired result.

**Problem 10.** Evaluate the expression $xz + zy$, knowing that $x$, $y$, and $z$ are positive solutions of the system of equations:

$$\begin{cases} x^2 + z^2 = 25, \\ y^2 + z^2 = 144, \\ z^2 = xy. \end{cases}$$

**Solution.** The conventional strategy to solve this problem is to find the numbers $x$, $y$, and $z$ as the algebraic solutions to the system and then plugging those values in to evaluate $xz + zy$. Whereas, as it will be demonstrated shortly, this is not the best decision in this case, we will show the steps in solving the system. This is a worthwhile exercise even from the standpoint of comparing two alternative ways for approaching this interesting problem.

This system of nonlinear equations can be solved by several different methods. We will concentrate on the method naturally derived from the given presentation of the equations. First, adding the first two equations and substituting the expression for $z^2$ in terms of $xy$ given in the third equation, we get

$$x^2 + 2z^2 + y^2 = 169,$$

$$x^2 + 2xy + y^2 = 169.$$

Observing that we obtained the perfect square on the left-hand side of the last equation and that all three given numbers are positive, it follows that

$$(x + y)^2 = 169,$$

from which $x + y = 13$. (1)

Next, subtracting the first equation from the second yields $y^2 - x^2 = 119$. Factoring the left side as the difference of squares and substituting 13 for $(x + y)$ from (1) gives $(y + x)(y - x) = 119$, or $(y - x) \cdot 13 = 119$, and finally,

$$y - x = \frac{119}{13}. \tag{2}$$

Combining (1) and (2), we now have to solve a simple system of two linear equations

$$\begin{cases} x + y = 13, \\ y - x = \frac{119}{13}. \end{cases}$$

Adding the equations, we find $y = \frac{144}{13}$. Substituting this into either of the equations gives $x = \frac{25}{13}$. Recalling the third equation from the original system $z^2 = xy$, we have

$$z = \sqrt{\frac{25}{13} \cdot \frac{144}{13}} = \frac{60}{13}.$$

The last step is to find
$xz + zy = \frac{25}{13} \cdot \frac{60}{13} + \frac{60}{13} \cdot \frac{144}{13} = \frac{60}{13}\left(\frac{25}{13} + \frac{144}{13}\right) = \frac{60}{13} \cdot 13 = 60.$

There is another approach, in which we will get to the desired outcome without calculating $x$, $y$, and $z$. The goal is to evaluate

$xz + zy$. Applying a geometrical interpretation of the given system of equations allows us to determine this sum without having to go through tedious calculations of each variable.

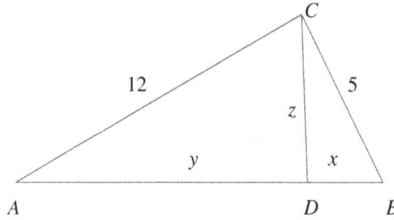

Let's consider two right triangles $BDC$ and $ADC$. The first with the legs $x$ and $z$ and the hypotenuse 5 (because it is given that $x^2 + z^2 = 25$ in the first equation of the system; by applying theorem converse of the Pythagorean theorem, such a right triangle has to exist) and the second right triangle with the legs $z$ and $y$ and the hypotenuse 12 (it has to exist since $y^2 + z^2 = 144$ from the second equation of the system).

Let's put them together, so they have the leg $CD$ as the common side. It follows that $CD$ is the altitude of the newly formed triangle $ACB$.

Recall now the third equation in the system, $z^2 = xy$.

Such a relationship between the altitude and the segments in which it splits the opposite side in a triangle is possible only in a right triangle. If readers are not familiar with this property, we hope they can easily prove it from the similarity of the triangles $ADC$ and $CDB$. Thus, $ACB$ has to be the right angle $(\angle C = 90°)$.

The area of the triangle $ABC$ equals the sum of the areas of the right triangles $BDC$ and $ADC$. The area of a triangle equals half of the base times the height dropped to that base. So, each area is determined as half the product of the legs:

$$S_{ABC} = S_{BDC} + S_{ADC} = \frac{1}{2}xz + \frac{1}{2}zy = \frac{1}{2}(xz + zy),$$

from which $xz + zy = 2S_{ABC}$.

On the other hand, the area of the right triangle $ABC$ equals

$$S_{ABC} = \frac{1}{2}AC \cdot CB = \frac{1}{2} \cdot 12 \cdot 5 = 30.$$

Therefore, $xz + zy = 2S_{ABC} = 2 \cdot 30 = 60.$

Comparing the two solutions, we learn that sometimes the conventional approach may not be the best one. Even though we are dealing with a system of equations, it was not the ultimate goal to solve it. It was required to evaluate $xz + zy$. A geometrical interpretation allowed us to skip several intermediate steps and proceed to the desired result directly.

**Problem 11.** This problem was offered by M. Akukov in magazine Квант (in Russian), #5, 2005 year.

Is it possible to find a triangle in which an altitude, a median, and an angle bisector drawn from different vertices intersect each other and form an equilateral triangle inside of the given triangle?

This intriguing and very interesting problem represents a different type of a challenge than we faced before. The question is whether such a triangle exists. If we manage to find even one such triangle, then our response will be positive. It will be natural to examine what conditions the original triangle has to satisfy so we will be able to form an equilateral triangle as requested in the problem. In fact, the simplest way to start our analysis is to assume that such a triangle indeed exists and investigate when this would be possible, if possible at all.

**Solution.**

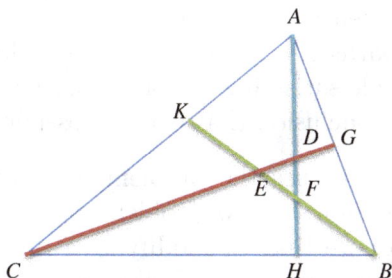

Consider triangle $ABC$ and draw $AH$, its altitude, $AH \perp BC$, $CG$, its median ($G$ is the midpoint of $AB$), and $BK$, its angle bisector, $\angle CBK = \angle ABK$. Assume that the points of their intersection $E$, $D$, and $F$ are the vertices of the regular triangle $EDF$. In a regular triangle, every angle equals $60°$, so $\angle EFD = 60°$. Angles $EFD$ and $BFH$ are vertical angles. Therefore, they are equal and $\angle BFH = \angle EFD = 60°$.

$AH \perp BC$, hence triangle $FHB$ is the right triangle in which $\angle FHB = 90°$ and $\angle BFH = 60°$. It follows that $\angle FBH = 90° - 60° = 30°$. Now, we can use the fact that $BK$ is an angle bisector of the angle $CBA$, and therefore, $\angle GBF = \angle HBF = 30°$. We obtained that in triangle $BEG$ there are two angles of 60° and 30°, $\angle GEB = 60°$ and $\angle GBE = 30°$. It implies that its third angle has to be a right angle:

$$\angle EGB = 180° - 60° - 30° = 90°.$$

Recalling that $G$ is the midpoint of $AB$, we conclude that $CG$ is the median and the altitude of the triangle $ABC$ at the same time. Therefore, $\triangle ACB$ is an isosceles triangle with $AC = BC$. As we proved before, $\angle GBF = \angle HBF = 30°$. So, $\angle CBA = 60°$, and we obtained that $\triangle ACB$ has to be an equilateral triangle. But, we know that in an equilateral triangle all its altitudes, medians, and angle bisectors drawn from different vertices coincide and intersect at one point, the center of $\triangle ACB$ (it is the orthocenter, centroid, and incenter of a triangle at the same time). This proves that it is impossible to get $\triangle EDF$; it degenerates into one point.

It merits pointing out that the question asked in the problem rendered the way we approached and contemplated the solution process. We started by making an assumption that the answer to the problem's question is positive. We did not make any attempts to prove or disprove the existence of an equilateral triangle. We worked out the problem backwards trying to investigate under what conditions an equilateral triangle satisfying the problem's conditions *may* exist, and arrived at the conclusion that it is impossible.

We are going to discuss the problems-relatives, converse problems, and the invention of new problems in subsequent chapters. It is interesting to note that by modifying a problem's details or a question asked, you may get a new challenge not even directly related to the solved problem. Alternately, in certain cases, by stating two seemingly unrelated questions and assuming you have to solve two different problems; in fact, you are solving conceptually exactly the same problem.

**Problem 12.** A tennis championship is played on a knock-out basis, i.e., a player is out of the tournament when he/she loses a match.

1. How many players participate in the tournament if 47 matches are totally played?
2. How many matches are played in the tournament if 60 players totally participate?

**Solution.** There are two different questions asked, so it looks like we have to solve two different problems. Let's see if that is the case here.

Consider the first problem and contemplate the plan for its solution from the very end — the final game played. Let's depict the problem in the table below. It will help to visualize it and better understand the question asked if we count the matches played backwards from the final match. See the column *Total matches played*:

| Match | Matches played at this stage | Total matches played | Total players |
|---|---|---|---|
| Final | 1 | 1 | 2 |
| Semifinals | 2 | $1 + 2 = 3$ | 4 |
| Quarterfinals | 4 | $1 + 2 + 4 = 7$ | 8 |
| $\frac{1}{16}$ Round stage | 8 | $1 + 2 + 4 + 8 = 15$ | 16 |

Obviously, half of the participants lose their matches at each stage, as it is seen in the table above, and are out of the tournament. The total number of matches played is always one less than the number of players participating in the tournament because every player, except the champion, has to lose one match. Therefore, there are a total of 48 players participating in the tournament with a total of 47 matches played.

We can clearly see now that the second question pondered is related to the same analysis of the problem. There will be a total of 59 matches played in the tournament with 60 participants. In fact, as we answered the first question, by applying the same reasoning, we would get the response to the second question as well. We indeed solved two different problems because they have different numerical values, but conceptually they are the same!

To conclude this chapter, we will explore one very compelling type of so-called "true-lie" problems that emphasizes the importance of asking key questions to get the proper result. These logical challenges are designed to get a solution depending on a very specific question to ask. There are also multiple appealing bypass variations flowing from these types of problems, analyses of which are closely related to each other.

**Problem 13.** A tourist stands at an unmarked intersection between the City of Lies and the City of Truth. Inhabitants of the City of Lies always lie. Inhabitants of the City of Truth always tell the truth. A citizen of one of those cities stands next to the tourist. What question should the tourist ask him to find his way to the City of Truth?

**Solution.** The question to ask should be "In which of those two cities do you live?" A liar will point to the City of Truth; a citizen of the City of Truth will also point in the direction to the City of Truth.

**Problem 14.** The teacher in a classroom is talking to identical twins and is trying to figure out who is who of the two of them. He knows that one kid, whose name is Brendon, is always telling the truth, and the other one, whose name is John, is always lying. Can he ask just one question to any of the two brothers to resolve his problem and find out who is who?

**Solution.** The question to ask can be "Is John a liar?"
If the answer is "Yes", then the kid, who gave it, is Brendon, so the other kid is John. If the answer is "No", then this response came from John, because he is a liar.

**Problem 15.** You are trapped in a room with two doors. One leads to a prison and the other leads to freedom. You don't know which is which. There are two robots guarding the doors. They will let you choose one door, but upon doing so you must walk through the door. You can, however, ask one robot one question. The problem is one robot always tells the truth, the other always lies (and both are well aware about each other's personality) and you don't know who is who. What is the question you need to ask in order to escape?

**Solution.** Ask one robot what the other robot would say if he was asked which door was safe. The liar will point to the door to prison (the robot who always tells the truth would respond with pointing to the door to freedom, so the liar has to point to the other door; he has to lie). The robot who tells the truth will point to the same door to the prison (because the liar answering the question about the safe door will point to the wrong door, the door to the prison, so the robot who tells the truth will have to repeat his answer). After getting a response, which will indicate the same door to the prison, go through the other door.

We hope that readers like and well understand the discussed problems because they would need to apply similar logic and analyses to be able to overcome the tricky challenges offered below. It worth remarking that many similar logical puzzles can be solved using the laws of Boolean algebra and logic truth tables, the discussion of which is beyond the scope of this book. Ambitious readers may wish to explore these interesting topics further.

Imagine that we visited an isolated island in an ocean with three types of inhabitants, Liars (they always lie), Truths (they always tell the truth), and Slyboots (sometimes they lie, sometimes they tell the truth; when they lie the first time, they tell the truth the second time). They all live in separate districts, but often visit each other.

Given the above information, solve the following problems.

**Problem 16.** While tourists were searching for a museum, they noticed a local guy standing by. They asked him who he was, trying to figure out first if they can trust him before asking for directions. "I am a Liar", was his answer. Can the tourists trust this guy and ask for directions?

**Problem 17.** Two local inhabitants, both of whom were not Slyboots (they all are at work at that time of the day), heard the conversation from Problem 16. One of them turned to the tourists and said, "My friend, unlike me, is a Truth". Can you tell who is who of these two men?

**Problem 18.** Perplexed by the previous two encounters, the tourists turned to another two people, who were passing by and asked them, "Are any of you Truths?" The first guy did not say a word, but the second person responded. After thinking for a few moments, the tourists figured out who was who. Can you? Before answering, recall that all Slyboots are at work at that time and neither of the two friends was a Slyboot.

**Problem 19.** One of the local inhabitants called an ambulance and said,

"We have a very sick relative in our family, please help".

"What part of the city do we have to go?" asked the ambulance driver.

"We live in the Slyboots' district", the caller said.

What district would the ambulance driver need to drive to?

**Problem 20.** Three of the island inhabitants from the different types (Liars, Truths, and Slyboots) decided to get recruited in the local army unit. All three of them were standing in line in front of the sergeant ready to answer his questions.

"Who is the second guy?" the sergeant asked turning to the first guy.

"He is a Liar", said the first person.

The sergeant turned to the second guy and repeated his question.

"I am a Slyboot", answered the second recruit.

The sergeant asked the third person the same question.

"He is a Truth", said the third guy.

Can you specify who is who among these three recruits?

In conclusion of this chapter, we offer one more interesting geometrical problem for the readers to try on their own. It has several different solutions. The one suggested in the "Solutions to Problems" section uses the Cartesian Coordinates method. The reason we demonstrated that specific solution is that it allows to go directly to the ultimate goal for finding the distance between two classic centers in a triangle (it is directed by the question asked). We suggest the readers find their own proofs and compare them. The more approaches you try, the better.

**Problem 21.** Given a right triangle with the legs 12 and 9, find the distance between its incenter (center of a circle inscribed in a triangle) and its centroid (the point of intersection of the medians).

# Chapter 4

# Thinking Outside the Box. Sophisms and Paradoxes

## 4.1 Playing Logical Games

An Attorney and a Mathematician are sitting next to each other on a flight from New York to Los Angeles. The Mathematician looks absentminded and very tired. The Attorney, who is bored, asks him to play a question-answers game with him. The Mathematician wants to relax and take a nap during the flight, so he politely declines. The Attorney decided to take advantage of his foolish-looking neighbor and then offered to play him for money under the following conditions: "I ask you a question and if you don't know the answer, you pay me only $5. Then you ask me a question, and if I don't know the answer, I will pay you $50". This catches the Mathematician's attention and to keep the Attorney quiet, he agrees to play the game.

"What is the distance from the Earth to the Moon in meters?" the Attorney asked. The Mathematician silently handed him $5. "Now it's your turn, please ask your question", the Attorney said.

"What climbs up a hill with three legs and comes down with four?" the Mathematician asked.

For the next 4.5 hours, the Attorney searches all the references he could find on the internet. He texts his all smart friends, but still to no avail. Finally, he wakes up his neighbor who was enjoying his sleep and gives him $50. The Mathematician pockets the $50 and goes right back to sleep. The Attorney is going nuts not knowing the answer. He wakes up the Mathematician and asks what the correct answer is. The Mathematician shrugs, reaches in his pocket, hands the Attorney $5 and goes back to sleep.

The moral of the story:

> Clearly understand a problem's data before starting to solve it.
> Pay attention to a problem's question or requirements; make sure it is clear and unambiguous for you.

Indeed, would the Attorney have slightly modified the game's rules and required a response to be provided by the asking party in case of an opponent's failure, he probably would not have lost that much money. On the other hand, the Mathematician, by analyzing the game's details, took full advantage of the game without even bothering to give a fair explanation to his neighbor. Thinking outside the box let him make $40 without any effort on his part and not even interrupting his sleep during the flight.

"Thinking outside the box" means to think creatively, freely, and unconventionally. This metaphor is often used in the business world. This is something that robots and smart machines can't do; it's what distinguishes humans from computers. Several years ago, I came across one intriguing story about marketing research results of investigating two similar candy store businesses located on neighboring streets in a small suburban town. Both stores sold a similar assortment of candies and cakes for teenagers. Even though the prices for comparable products were the same, one store was very profitable and attracted most of the kids, while the other one barely made ends meet. So, there is a problem:

> There are two similar stores offering the same products for the same price. Find some logical explanations for what is done differently by the store owners to attract/deter the customers to arrive at such opposite profit results.

I suggest the readers think about their versions of the possible solutions before reading the marketing team's amazing findings.

In the profitable store, the salesman taking the order from the kids put a small number of candies on a weighing scale and then added additional candies to get to the ordered number of pounds by the customer. In the other store, the salesperson did the opposite thing. For example, when a kid asked to weigh him 2 pounds of candies, a salesperson put a little more than the requested 2 pounds and took candies out from the weighing scale until the weight was exactly 2 pounds. In the first store, in this example, the salesperson had originally less than 2 pounds on the scale and then added candies to get to the ordered 2 pounds. Observing these practices, the marketing team concluded that kids did not like seeing candies being taken out from the scale. The perception was that the salesperson took away their candies (which are not theirs yet!), while in the profitable store kids liked the idea of added candies; they felt like they were getting more. Through the analysis of a specific behavior, reaction, and feedback from the customers, along with a little creativity, this provided great financial outcome for the business owner of the first store!

The results from various psychological studies reveal that in most cases when we introduce a goal in our thinking, we get a constraint as well. Our mind now has a direction, and it will tend to go in that direction. This is why many businesses bring in outside consultants to help them come up with new ideas. The outsiders don't carry the burden of constraints on their thinking. They can generate new ideas and lead to innovative accomplishments.

Some very interesting results were obtained in the 2014 University of Colorado Boulder studies of children behavior: the more time children spent in less structured activities, the better their self-directed executive function. Conversely, the more time children spent in more structured activities, the poorer their self-directed executive function. So, the goal is to find such activities that help enhance visualizing the problem, making it interesting and non-threatening. Is it possible to teach thinking outside the box? I believe it can be done through playing cute logical games, examining pathfinding from labyrinths, and solving paradoxes and unorthodox puzzles that wake up your creativity and imagination. These activities are very useful in developing and stimulating problem-solving skills as well.

**Game 1.** There are two piles of peanuts on the table, 24 in one and 19 peanuts in the other. A brother and sister like the peanuts and invented the following game: taking turns, each is allowed to eat all the peanuts in a pile while dividing the peanuts in the other pile into two new piles. There are no restrictions on the number of peanuts allowed in the new piles, but whoever would not be able to divide the remaining peanuts in two piles because there is only one peanut left, loses the game. What is the best strategy for the sister to play the game and win having the first turn?

**Solution.** Through analyzing the data of the problem, to invent the winning strategy, we need to replay the game in a reverse order. The loser would have one peanut remaining to eat. Indeed, the last possible division is related to the pile consisting of three peanuts.

If the sister manages to play the game in such a way that she would have her turn having three peanuts remaining, she would win. She would eat one peanut and divide the other peanuts into two piles with one peanut in each. By eating one peanut, the brother would have only one peanut remaining which is impossible to divide. Now, we see that when making the first turn, the sister has to eat all 19 peanuts and divide 24 peanuts into piles consisting from 1 and 23 peanuts. If the brother eats 23 peanuts, he immediately loses the game. So, the brother has no choice other than to eat 1 peanut and divide 23 peanuts into two piles. One of the new piles would have an even number of peanuts and the other pile would have an odd number of peanuts. The sister has to eat all the peanuts in the pile with an odd number of peanuts and divide the second pile into the one-peanut pile and the pile with an odd number of peanuts. By doing this, she forces her brother to play under her rules repeating similar steps all the way until they get to the point when there are only three peanuts remaining.

**Game 2.** In a summer camp, 68 kids are playing the following game:

They form one big circle; each kid having assigned a number from 1 to 68, takes his respective position in the circle. Every other kid starting from the second number has to step out from the circle. They repeat this till only one person remains in his original position taken on the circle. That kid is the winner. As they play, the counting starts from the first person and goes in a clockwise direction. What starting position do you need to take to win the game?

**Solution.** Assuming the number of kids playing this game would have been $2^n$, after the first round eliminating half of the players, we would return to the kid standing at the first place. This will be repeated $n$ times. In such a scenario, the winner would be the kid originally standing at the first spot. However, there are 68 players, which is not a power of 2. The closest power of 2 to this number is $2^6 = 64$. The difference is $68 - 64 = 4$. By eliminating the players under the numbers 2, 4, 6, and 8 in the first round, we are theoretically left with 64 players and the count starts from the kid standing at the ninth spot.

As we have already discussed, in this case, after six eliminations, the kid at the ninth spot will become the winner.

**Game 3.** Two players taking turns move the hour hand of a clock by 2 or 3 hours forward. The game starts when the hour hand is at 12. The winner of the game is the player who would be able to place the hour hand at 6. What is the playing strategy to win?

**Solution.** When making the first move, the first player has to place the hour hand at 2, because by placing it at 3, he would allow his opponent to make his move to 6, and he would lose the game. Similarly, the second player in his move has the only choice to place the hour hand to 5; if he puts it at 4, he would lose the game allowing the first player to place it at 6. Next, the first player has to move the hour hand to 8, because no matter what his opponent would do making his next move either to 10 or to 11, the first player would need to place the hour hand at 1 (if the second player goes to 10, the first player moves the hour hand by 3 hours, if the opponent moves to 11, he moves the hour hand by 2 hours). As soon as the first player gets to 1, he is in the position to win the whole game. Indeed, if his opponent moves by 3 hours and gets to 4, then the first player will move the hour hand by 2 hours and gets to 6; if the second player goes to 3, then the first guy will move the hour hand by 3 hours and still win the game.

Let's play another logical game. Consider the following chain of cute riddles-jokes. "Chain" here means that the riddles below are related in one way or another. Solving each, you somehow rely on the previous one(s). "Somehow" in the last sentence is up to you; you decide how to use the given information, and how every following problem is related to the preceding ones.

**Game 4.**

**Problem 1.** How do you put an elephant in a cupboard?

**Problem 2.** How do you put a giraffe in a cupboard?

**Problem 3.** A lion, the king of animals in the jungle, ordered all the animals to get together for an annual meeting in order to discuss important matters. Everybody obeyed the command except one animal. Who was absent during that meeting?

**Problem 4.** A group of visiting tourists is standing at the banks of a river full of crocodiles. They have two small boats with enough space to accommodate all the group's members. Is it safe to use those boats to cross the river, and not fall subject to the crocodiles' attack?

**Solutions:**

**Solution to Problem 1.** Open the cupboard. Put an elephant inside. Close it.

**Solution to Problem 2.** Open the cupboard. Take the elephant out. Put a giraffe inside. Close it.

**Solution to Problem 3.** The giraffe was absent. He was in the cupboard.

**Solution to Problem 4.** Yes, it is safe. The tourists don't have to be concerned about a possible crocodiles' attack. All the crocodiles are at the annual animals' meeting called by the lion king. There are no crocodiles in the river at the moment, so it is safe to cross, even in small boats.

I find it amazing how easily and I would say, naturally, small kids approach many problems that seem difficult for older kids or adults. They think without any restrictions on commonly accepted or assumed things. When solving the first riddle, an adult would face the impossible task on fitting an elephant into cupboard, assuming that it is a real animal. For a kid, it's just a game. Nobody told you in the problem's statement that you are dealing with a real animal, not a toy. So, it is possible to solve the problem as suggested. The second problem looks much easier as soon as you realize how to solve the first one. An adult would be completely at sea by the third problem. On the other hand, kids that accept the rules of the game (not even announced rules!) would provide an obvious response to the third question — the giraffe is in a cupboard (see Problem 2), so how can he be at the meeting? A similar approach works while solving Problem 4.

Some skeptics would argue that Problems 3 and 4 are not related to Problems 1 and 2 in real life. Well, this is a game, as it was warned before. Don't you agree there is a logical connection if we follow those unannounced but pretty obvious "rules" of the game?

This appealing game reveals some interesting features about the "thinking outside the box" approach. Sometimes, an easy solution could be accomplished by directly following the problem's details and utilizing them, even when it looks strange and absurd at first glance. Observe where the selected path takes you, establish logical links, justify your steps, and see if you can get to the desired destination. And, use the previously solved problems, especially the ones related to your problem. Those family connections prove invaluable on many occasions.

I believe that playing similar logical games to the ones illustrated above, studying various paradoxes and puzzles unexplainable at first glance, is a great tool for boosting creativity and imagination, as well as thinking out of the box abilities. They always wake up an interest in math and stimulate a desire to figure out the trick or paradox without using any outside help. Starting from simple ones and then proceeding to more and more complicated problems, you master problem-solving techniques, and develop a deeper understanding of the subject matters involved.

## 4.2   Sophisms

Various diagrams, additional constructions, and pictures in many cases simplify a problem's solution; they are useful in designing a plan for a solution, making the problem manageable. However, there is an opposite tendency to completely rely on a picture, assuming it gives some clear-cut information in the solution process. Even worse, one may substitute a wrong but appealing and preferable assumption fitting his/her notion of the facts based on some misleading picture. How many times a math teacher hears the following argument from a student: "Why do I have to prove that these lines are parallel, (or segments equal, angles have the same measure, etc.), if I see it on the picture?"

We sometimes accept certain "vivid" pictures and plausible but deceptive arguments as valid facts and use them in logical justifications of intermediate statements or in a problem solution process arriving eventually at wrong conclusions. One erroneous assumption or incorrect guess regarding the elements' locations or values (segments lengths, angles values, areas, volumes, etc.) should ruin the whole logical structure and lead to a false result.

Look carefully at the five figures below.

Do you believe your eyes?

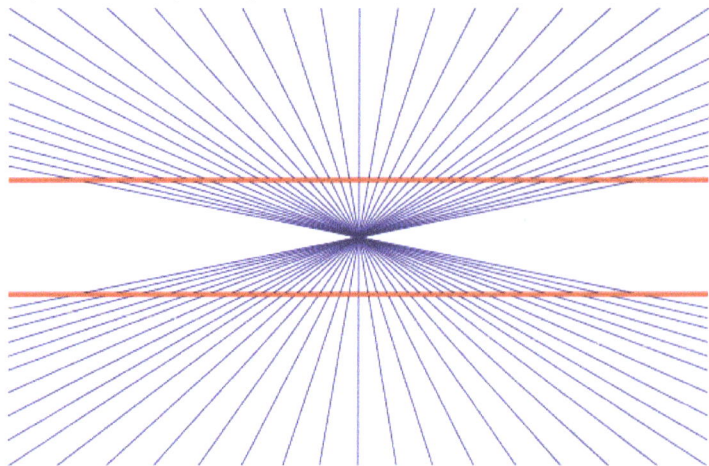

(1)

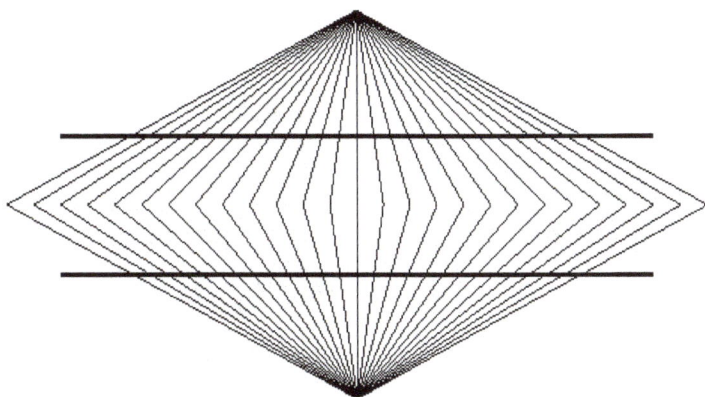

(2)

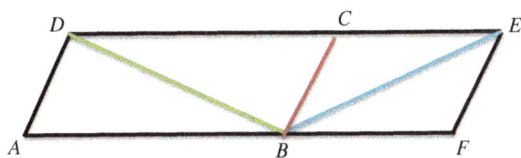

(3)

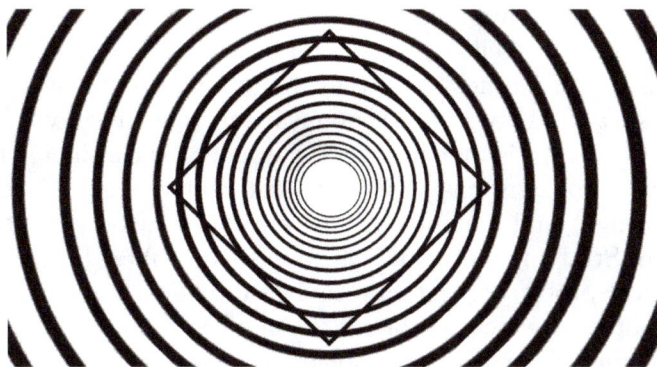

(4)

(5)

- Don't you feel like the two middle lines in Figure 1 curved upward and the two middle lines in Figure 2 are curved inward?
- What about comparing the lengths of segments $BD$ and $BE$ in Figure 3? Doesn't it look like $BD > BE$?
- Multiple circles in Figure 4 contribute to viewing the sides of the square as slightly curved inward? Don't they?
- Finally, which of the two segments in Figure 5 has the greatest length? It seems like the one located to the left, doesn't it?

In reality, however, all four lines in Figures 1 and 2 are parallel, $BD = BE$ in Figure 3, the rectangle's sides are straight line segments not curved at all in Figure 4, and both segments in Figure 5 are of the same length!

Logical and mathematical sophisms (fallacious argument, especially one used deliberately to deceive) and paradoxes are very useful in developing critical thinking, inventiveness; they serve as a great practical tool in reinforcing deep understanding and verification of every step in a problem-solving process. Sophisms played a huge role in the history of mathematics; they stimulated the discovery of new mathematical principles and theorems in many instances when mathematicians found mistakes in proofs that had been considered flawless for years. A sophism has to be identified (sometimes, it becomes a real challenge) and spelled out; the error has to be corrected and explained. Studying mathematical sophisms justifies the necessity of rigid logical proofs, relying on established axioms and theorems. In the case of geometrical problems, it emphasizes the importance of not reasoning using figures. Instead, one must verify for accuracy and correctness all geometrical drawings.

So, let's continue with our fun games that we started at the beginning of the chapter. We invite the readers to identify errors made in the following mathematical sophisms and explain the paradoxes.

**Problem 5.** Let's make several modifications of the equality

$$225 : 25 + 75 = 100 - 16$$

and prove that $5 \cdot 5 = 7$.

Indeed, 225, 25, and 75 have a common factor of 25. Factoring the left side of the given equality yields

$$25 \cdot (9 : 1 + 3) = 25 \cdot (9 + 3) = 25 \cdot 12.$$

We have on the right-hand side of the original equality

$$100 - 16 = 84 = 7 \cdot 12.$$

Comparing the results, we see that $25 \cdot 12 = 7 \cdot 12$, which leads to $5 \cdot 5 = 7$. What went wrong in our modifications?

**Problem 6.** Often, when we want to express that something is very obvious to us, we say, "This is as clear as $2 \times 2$ is $4$". Is it really?

It is given that for some real numbers $x$, $y$, and $z$, they relate as $x - y = z$. Multiplying both sides by 5 and by 4, we get two slightly modified equalities $5x = 5y + 5z$ and $4y + 4z = 4x$. Adding these equalities gives

$$5x + 4y + 4z = 5y + 5z + 4x,$$

Subtracting $9x$ from both sides leads to $4y + 4z - 4x = 5y + 5z - 5x$ or equivalently, $4 \cdot (y + z - x) = 5 \cdot (y + z - x)$. Canceling out a common factor of $(y + z - x)$, we obtain that $4 = 5$. So, is it really $2 \cdot 2 = 4$?

**Problem 7.** Find the error in the following calculations.

We start with the correct equality $4 - 10 = 9 - 15$.

Adding $6\frac{1}{4}$ to both sides gives $4 - 10 + 6\frac{1}{4} = 9 - 15 + 6\frac{1}{4}$.

Rewriting the last equality using the identity for the square of a difference $a^2 - 2ab + b^2 = (a - b)^2$ gives

$$2^2 - 2 \cdot 2 \cdot \frac{5}{2} + \left(\frac{5}{2}\right)^2 = 3^2 - 2 \cdot 3 \cdot \frac{5}{2} + \left(\frac{5}{2}\right)^2,$$

or equivalently, $(2 - \frac{5}{2})^2 = \left(3 - \frac{5}{2}\right)^2$. This leads to $2 - \frac{5}{2} = 3 - \frac{5}{2}$, from which $2 = 3$.

**Problem 8.** Evaluate the infinite sum $1 - 1 + 1 - 1 + 1 - 1 + \cdots$.

In order to find the value of this infinite expression, let's denote it by $S$,

$$S = 1 - 1 + 1 - 1 + 1 - 1 + \cdots .$$

We can rewrite it as $S = 1 - (1 - 1 + 1 - 1 + 1 - \cdots)$.

Clearly, the expression in parenthesis is $S$ as well, so the last equality modifies to $S = 1 - S$, from which $S = \frac{1}{2}$.

How does the obtained result agree with the fact that

$$\underbrace{1 - 1 + 1 - 1 + 1 - 1 + \cdots}_{n \text{ addends } (n \to \infty)} = \begin{cases} 0, & \text{if } n \text{ is even} \\ 1, & \text{if } n \text{ is odd} \end{cases} ?$$

**Problem 9.** Given the right triangle $ABC$ ($\angle C = 90°$) and $D_1$ is the midpoint of $AB$, perpendiculars $D_1M_1$ and $D_1N_1$ are dropped from $D_1$ to $AC$ and $BC$, respectively. New right triangles $AM_1D_1$ and $D_1N_1B$ are formed. Repeat the similar constructions two more times for $D_2$ and $D_3$, the midpoints of $AD_1$ and $D_1B$.

We got the broken line $BN_3D_3M_3D_1N_2D_2M_2A$ the length of which equals the sum of legs $AC$ and $BC$.

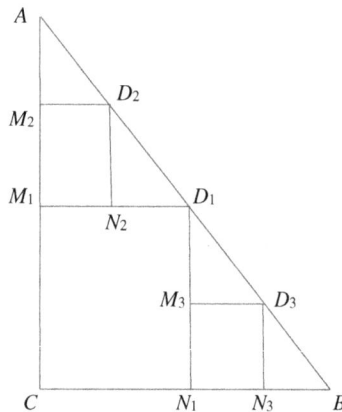

Assume we extend this process of dropping the perpendiculars infinitely many times. Then, the length of the broken line will be more and more closely approaching the hypotenuse $AB$, while its length still will be equal to $AC + BC$, the sum of the legs. Didn't we just prove that in the right triangle $ABC$ the length of its hypotenuse equals the sum of its legs, $AC + BC = AB$? How does this agree with the Pythagorean Theorem that $AB^2 = AC^2 + BC^2$?

In the following problem, we will prove an even more powerful statement that in a right triangle a leg has the same length as the hypotenuse.

**Problem 10.**

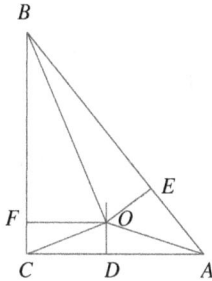

Consider the right triangle $ABC$ ($\angle C = 90°$) and draw its angle bisector of angle $B$. Next, draw a perpendicular bisector of $AC$ through its midpoint $D$ till its intersection with the angle bisector of angle $B$ at $O$. Connect $O$ with $A$ and $C$ and drop perpendiculars $OE$ and $OF$ to $AB$ and $BC$, respectively. Let's prove that $BC = BA$.

Since $BO$ is an angle bisector of angle $B, \angle FBO = \angle EBO$. Right triangles $BFO$ and $BEO$ are congruent by angle-side-angle (ASA): $BO$ is the common hypotenuse, $\angle FBO = \angle EBO$, and $\angle FOB = \angle EOB$, as the differences of 90° with equal respective acute angles. Therefore, $FO = EO$ and $BF = BE$. By construction, $DO$ is a perpendicular bisector to $AC$. Hence, the right triangles $CDO$ and $ADO$ are congruent by side-angle-side (SAS): $CD = AD$ and $OD$ is the common leg. So, we see that $CO = AO$. Recalling that $OE = OF$, we can now conclude that right triangles $OFC$ and $OEA$ are congruent as well, from which it follows that $CF = AE$.

Considering that $BC = BF + FC$ and $BA = BE + EA$, we obtain that the right-hand sides of both equalities are equal. Therefore, $BC = BA$.

Can you find an error in the above proof?

**Problem 11.** Explain what went wrong with the following modifications of the well-known trigonometric *Pythagorean Identity* $\sin^2 \alpha + \cos^2 \alpha = 1$.

Expressing one of the trigonometric functions in terms of the other, we have $\cos^2 \alpha = 1 - \sin^2 \alpha$. Raising both sides to the power $\frac{3}{2}$ gives $(\cos^2 \alpha)^{\frac{3}{2}} = (1 - \sin^2 \alpha)^{\frac{3}{2}}$, which simplifies to $\cos^3 \alpha = (1 - \sin^2 \alpha)^{\frac{3}{2}}$. Now, adding 1 to both sides and squaring the obtained sums yields

$$(\cos^3 \alpha + 1)^2 = ((1 - \sin^2 \alpha)^{\frac{3}{2}} + 1)^2.$$

Let's evaluate the last equality substituting for $\alpha$ values $\frac{\pi}{2}$ and $\pi$:

If $\alpha = \frac{\pi}{2}$, then $(\cos^3 \frac{\pi}{2} + 1)^2 = (0 + 1)^2 = 1$;
$((1 - \sin^2 \frac{\pi}{2})^{\frac{3}{2}} + 1)^2 = (0 + 1)^2 = 1$.

So, we get $1 = 1$, which is a correct equality.

However, if $\alpha = \pi$, then $(\cos^3 \pi + 1)^2 = (-1 + 1)^2 = 0$, while $((1 - \sin^2 \pi)^{\frac{3}{2}} + 1)^2 = ((1 - 0)^{\frac{3}{2}} + 1)^2 = (1 + 1)^2 = 4$, and we get that $0 = 4$, which is absurd.

How come did we get different numerical values when we substituted $\frac{\pi}{2}$ and $\pi$ into the valid equality?

**Problem 12.** Can you explain what is wrong in the following manipulations?

We start with the obvious inequality $\frac{1}{4} > \frac{1}{8}$. It implies that $(\frac{1}{2})^2 > (\frac{1}{2})^3$. Taking common logarithm of both sides gives $\log \left(\frac{1}{2}\right)^2 > \log(\frac{1}{2})^3$, or equivalently $2 \log \frac{1}{2} > 3 \log \frac{1}{2}$. Canceling out $\log \frac{1}{2}$ on both sides of the last inequality, we obtain that $2 > 3$. How is this possible?

**Problem 13.** Given $\sqrt{45 - x} + \sqrt{9 - x} = 4$, find $\sqrt{45 - x} - \sqrt{9 - x}$.

**Solution.** To solve the problem, we will use the formula for the difference of squares

$$a^2 - b^2 = (a - b)(a + b).$$

Denoting $\sqrt{45 - x} = a$ and $\sqrt{9 - x} = b$, we have $a^2 - b^2 = 45 - x - 9 + x = 36$.

So, now we can find $(a - b)$ as $a - b = \frac{a^2 - b^2}{a + b} = \frac{36}{4} = 9$.

We got a very interesting result. Having that $a + b = 4$, we obtained that $a - b = 9$.

By definition, the square root of a number $y$ is always a non-negative number, $\sqrt{y} \geq 0$. Hence, $\sqrt{45 - x} \geq 0$ and $\sqrt{9 - x} \geq 0$. How come the difference of two non-negative numbers is greater than their sum? What was wrong with our solution?

**Problem 14.** The numbers $b$ and $c$ satisfy the equation $x^2 + bx + c = 0$. Find $b$ and $c$.

**Solution.** Applying Viète's Formulas for the roots of a quadratic equation, we have the following system:

$$\begin{cases} x_1 + x_2 = -b, \\ x_1 \cdot x_2 = c \end{cases}$$

or equivalently, because $b$ and $c$ satisfy our equation,

$$\begin{cases} b + c = -b, \\ b \cdot c = c. \end{cases}$$

Solving this system, we easily get $b = c = 0$ and $b = 1$, $c = -2$. Do you have any concerns about this solution? Did we solve the problem?

## 4.3 Paradoxes

A paradox is a statement that despite valid reasoning from true premises leads to an apparently self-contradictory or logically unacceptable conclusion. Some paradoxes can be explained and unfolded; some logical paradoxes are known to be invalid arguments; some paradoxes, as *Russell's paradox*, for example (it questions whether a "set of all sets that do not contain themselves" would include itself), have revealed errors in definitions assumed to be rigorous, and have caused some mathematical and logical statements to be re-evaluated. Russell's paradox (after British mathematician and philosopher Bertrand Russel (1872–1970)) has a vivid interpretation in the so-called *Barber paradox*:

> *We define the barber as the person who shaves all those, and those only, who do not shave themselves. Does the barber shave himself?*

As simple as it looks, answering this question, though, results in a contradiction! Indeed, the barber shaves only those who do not shave themselves, so he cannot shave himself by definition of being a barber. On the other hand, if the barber does not shave himself, then he is one of those who would be shaved by the barber, and thus, as the barber, he must shave himself.

There are also paradoxes such as *Curry's paradox* (after American mathematician and logician Haskell Brooks Curry (1900–1982)) that uses a particular kind of self-referential conditional sentence. It has also been called *Löb's paradox* after German mathematician Martin Hugo Löb (1921–2006). The paradox relies on one accepting that some sentence $A$ is "true". An arbitrary claim $B$ is proved from the mere existence of a sentence $A$ that says of itself "If $A$, then $B$", requiring only a few banal logical deduction rules. Since $A$ is arbitrary, any logic having these rules proves everything. It occurs in naive set theory, and allows the derivation of an arbitrary sentence from a self-referring sentence. This paradox demonstrates the inconsistency of certain systems of formal logic.

The detailed discussion of this interesting paradox is beyond the scope of our book. We suggest that the ambitious readers find more about it in mathematical literature. One such source, for instance, is Haskell B Curry, "The paradox of Kleene and Rosser", *Transactions of the American Mathematical Society*. 50.3 (1941): 454–516.

I strongly believe that studying various paradoxes is worthwhile and useful in awakening interest in mental activities and developing critical thinking.

We offer below for your amusement two geometrical paradoxes.

The first is one of the most famous "dissection" paradoxes known as *Curry's paradox* after amateur magician Paul Curry (1917–1986). It should not be confused with Haskell Curry's paradox mentioned above, and neither has anything to do with *Golden States Warriors* Stephen Curry's magical basketball paradoxes!

By cutting a plane geometrical figure into several parts and then rearranging the parts, it's possible to get a figure of a different shape, but the area should not change; both figures have to be of the same area. This is not the case though in the following Problem 15. Can you explain it?

**Problem 15.** Square versus rectangle area paradox.

Two figures are composed of the same shapes, but with different areas.

Examine the square and the rectangle below. They are composed of exactly the same set of shapes, two right triangles with legs 3 and 8, and two congruent right trapezoids with bases 3 and 5. However, the area of a square equals $8 \times 8 = 64$ square units, while the area of a rectangle equals $5 \times 13 = 65$ square units. How is this possible?

Can you give any other similar examples selecting different numerical values for a square's side? How is this paradox related to the properties of the Fibonacci numbers?

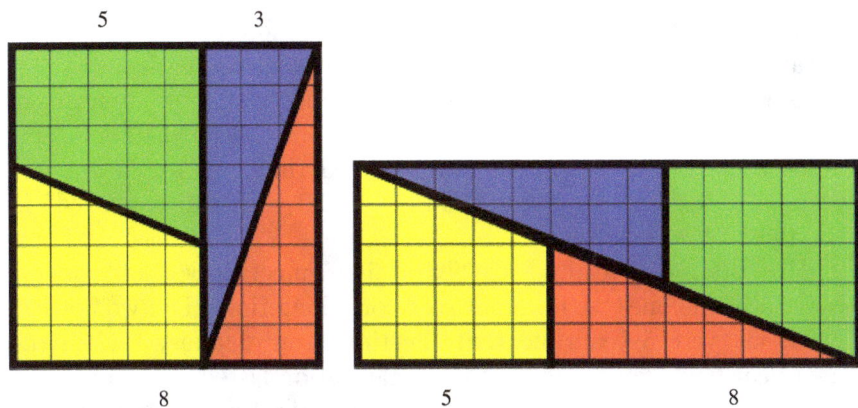

**Problem 16.** "Vanished" segment paradox.

Draw 13 segments inside the rectangle and then cut along the straight line connecting the upper end of the first and the bottom end of the last segment (see figure at left). Slide the top piece to the left one line (see figure at right). Count the segments. You have now 12 instead of 13. Can you explain what happened to one of the segments? Where did it go?

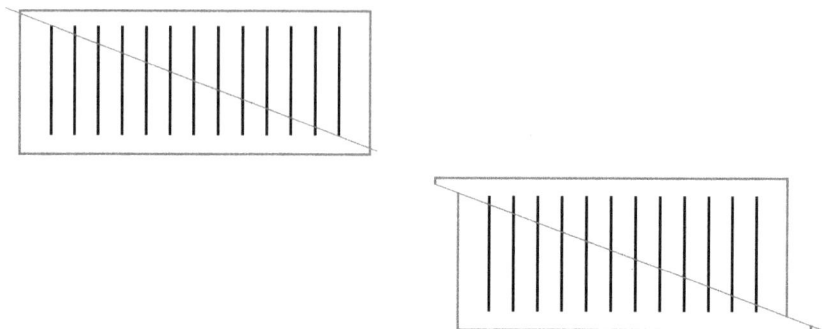

Explanations to both geometrical mysteries, as well as the answers to all the above sophisms, are given at the end of the book in the "Solution to Problems and Answers section". We hope, though, that the readers will try to figure out these tricks and have fun in doing so before checking the answers.

To conclude this chapter, we explore one more paradox, an anecdotal explanation of which with the help of mathematical calculations I find cute and appealing at the same time. I got it from an old friend of mine when we were recalling our college years and how quickly life passes us by. We both agreed that our 4 college years seemed to last forever, while the next 10–20 years "disappeared" in the blink of an eye.

How often did we allude to "flying time" meeting with old friends, especially the ones we have not seen for a long time? Did we not feel when we were 5 years old that the time from January 1st till the next year Christmas and New Year celebration lasted very, very, very long? The same story with waiting for our next Birthdays; it felt like it takes a lifetime from the seventh to eighth birthday. As we are getting older, the common perception is that time takes acceleration and Birthdays along with New Year's celebrations fly one after the other with increasing speed. Is it really? Why do we feel that way? Is there any explanation for this paradox?

German philosopher, physicist, and experimental psychologist Gustav Theodor Fechner (1801–1887) formulated the so-called Weber–Fechner Law, establishing the nonlinear relationship between psychological sensation and the physical intensity of a stimulus. Fechner and his teacher, a German physician Ernst Heinrich Weber (1795–1878), performed extensive studies of how humans perceive physical magnitudes. According to Fechner's law, human perceptions of sight and sound work as follows:

Perceived loudness/brightness is proportional to logarithm of the actual intensity measured with an accurate non-human instrument, $p = k \ln \frac{S}{S_0}$, where $S$ is the value of the intensity, $S_0$ is the lower bound value of intensity perceived (if $S < S_0$, then we do not perceive any difference), and $k$ is some constant dependent on the subject of perception.

If we have in our left hand a weight of 100 grams and in our right hand a weight of 110 grams, the difference in weights will hardly be noticeable. But, if the mass is substantially increased (doubled, for instance), then the difference will become noticeable. A similar example can be provided with vision perception of the brightness of a subject. If we have 60 lighting candles and add one candle, we would not feel the same difference when we add one candle to 120 lighting candles. We would need to add two candles because the difference in brightness becomes noticeable with a change of around 1.2%.

- 1.2% of 60 is 0.72, which means we need one candle;
- 1.2% of 120 is 1.44 (close to 2), so we need two candles.

What about if we apply the same reasoning to time perception changes for different ages? For a 10-year-old kid, 1 year represents $\frac{1}{10}$ of his life, while for a 60-year-old person, 1 year is $\frac{1}{60}$ of his life, a much smaller fraction. Considering our life as some constant (sure, nobody knows his/her exact number, but still it is some fixed constant), we should have different perceptions of time periods depending on the age and fraction/percent calculation. As we said, for a 10-year-old kid, 1 year constitutes 10% of his life and is regarded as some substantial time period; at 25 years, 1 year is 4% of the total and is perceived differently; at 50 years, 1 year is only 2% of the total. No wonder, with aging, we feel that a 1-year time interval becomes smaller and is apprehended as evaporating much faster than at the younger ages. In other words, the longer we live, the shorter each passing year is perceived in relation to our whole lives. Keeping this in mind, we should enjoy every fun moment today, because tomorrow these moments become shorter!

Chapter 5

# "Precise Steps" Problems. Playing Preferans

There is one compelling type of problems, called "river crossing puzzles", in which very precise steps are to be taken in a specific sequence to accomplish the goal of transporting people and goods from one bank of a river to the other bank. The difficulty of the puzzle depends on the restrictions on which or how many items can be transported at the same time, or which or how many items may be safely left together. One of such problems is the old well-known riddle about a farmer crossing a river and taking with him a wolf, a goat, and a cabbage.

**Problem 1.** There is a boat that can fit a farmer and either the wolf, the goat, or the cabbage. If the goat and cabbage are left alone, the goat will eat the cabbage. If the wolf and the goat are left alone, the wolf will eat the goat. Only when the farmer is present are the goat and cabbage safe from their enemies. How can the farmer carry the wolf, the goat, and the cabbage across the river?

**Solution.** The farmer has to do the following in the indicated sequence:

- He takes the goat across. The wolf and cabbage are left behind.
- The farmer returns to the bank of the river where he left the wolf and the cabbage.
- He takes the wolf across.
- He returns with the goat to pick up the cabbage, leaving the wolf behind alone.

- He picks up the cabbage and leaves the goat behind.
- The farmer returns to take the goat (the wolf and the cabbage are already on the other bank of the river).
- Finally, the farmer takes the goat across.

As we can see, each step has to be well thought and all the steps are to be arranged in a specific way to accomplish the goal. Even a slight error will result in a disaster; either the cabbage or the goat might be eaten.

Let's go over another more complicated example.

**Problem 2.** Once upon a time, three knights along with their servants approached the river and wanted to cross it. They found a tiny boat that can only hold two people. The problem became a little more complicated because every knight was reluctant to leave his servant with the other knights. They all insisted to have each servant next to his knight. How did they manage to cross the river?

**Solution.** Let $A$, $B$, and $C$ be the three knights and $a, b$, and $c$ represent their respective servants. We will illustrate all our steps in crossing the river with two tables representing each bank of the river. Original disposition:

| Left bank | Right bank |
|:---:|:---:|
| $ABC$ | |
| $abc$ | |

**Step 1.** Any two servants cross the river

| Left bank | Right bank |
|:---:|:---:|
| $ABC$ | |
| $c$ | $ab$ |

**Step 2.** One of the servants returns to pick the third servant

| Left bank | Right bank |
|:---:|:---:|
| $ABC$ | |
| | $abc$ |

**Step 3.** One of the servants returns to stay with his knight while the other two knights cross the river

| Left bank | Right bank |
|:---:|:---:|
| $C$ | $AB$ |
| $c$ | $ab$ |

**Step 4.** One of the knights ($B$, for example) returns along with his servant, leaves him, and takes the remaining knight to the other bank

| Left bank | Right bank |
|:---:|:---:|
|  | $ABC$ |
| $cb$ | $a$ |

**Step 5.** The servant $a$ returns to take one of the other two servants, servant $b$, for example,

| Left bank | Right bank |
|:---:|:---:|
|  | $ABC$ |
| $c$ | $ab$ |

**Step 6.** Knight $C$ returns to pick his servant and crosses the river with him completing the crossing

| Left bank | Right bank |
|:---:|:---:|
|  | $ABC$ |
|  | $abc$ |

Once again, only the precise execution of each step allowed to solve the problem under the given conditions and restrictions.

Let's consider one more problem, not belonging to "river crossing puzzles" but conceptually similar in nature.

**Problem 3.** There are 12 volumes of Britannica encyclopedia standing in the following order on the book shelf: 1, 2, 3, 4, 5, 6, 7, 8, 9, 10, 12, 11 (the last two volumes are not in order). You are allowed to take out and then to insert to a different place any three consecutive volumes on that shelf. Is it possible to get all the volumes in order after several such rearrangements?

**Solution.** This problem is similar to the discussed problems because it requires a very specific set of steps to be taken to achieve the desired result. You should be able to succeed in putting all the volumes in order, for example, by the following three rearrangements:

1, 2, 3, 4, 5, 6, 7, 8, 9, 10, 12, 11

1, 2, 3, 4, 5, 6, 7, 12, 8, 9, 10, 11

1, 2, 3, 4, 5, 6, 7, 10, 11, 12, 8, 9

1, 2, 3, 4, 5, 6, 7, 8, 9, 10, 11, 12

There are many more interesting problems similar to the ones discussed above. I hope the readers get a general idea about the nature of these challenges. We suggest now to get acquainted with one fascinating intellectual card game based on the similar but much more advanced idea of determining the proper sequence and order of steps to get to the desired result.

In many mathematical problem-solving books, chess etudes, compositions, and various chess-related problems are included as engrossing logical challenges. I found this as a great resource for additional practice in mental operations, designing a plan for a solution (foreseeing possible steps for winning combination), developing logical thinking and creativity. There is another amazing game, a card game called preferans, which genuinely competes with chess in achieving these goals. Even though it is a card game, preferans belongs to a category of so-called "commercial" games distinguished from the gambling games, meaning that the final outcome significantly depends on the skill level of a player rather than pure luck. In preferans, money is not a strategic element, thus making it a non-essential part of the game. The game is popular in the republics of the former Soviet Union and in several Eastern European countries, while it is relatively unknown in the rest of the world. In my opinion, this is the most exciting and challenging of all intellectual games; it compares even with such an established leader as chess. Noteworthy, many great chess masters of the past and present times, among which there are several world champions, love preferans and often played and play it. In chess, you play a prolonged game strategically building your way to a victory through the entire game. On the contrary, when playing preferans, you theoretically do not have any single game that is more or less important than the other one. You play till you get a specific number of winning points with every game providing a unique challenge for the participants. Some games are routine, some are more interesting, and some present real challenges to get to the desired result. The unpredictability of each and every round and your ability (skills) as a player make the game so unique and compelling. Speaking about the possible number of combinations, Polish author S. Jeleński in his book *Śladami Pitagorasa — Rozrywki matematyczne*, (in Polish) Poznań, Drukarnia Św. Wojciecha provided the following calculation:

$$\frac{32!}{10! \times 10! \times 10! \times 2!} = 2,753,264,408,504,640,$$

where $n!$ is the product of all numbers from 1 to $n$.

Indeed, it is a great source for practicing mental operations and developing logical thinking and problem-solving abilities.

Preferans is normally played by three or four players with a 32-card piquet deck with the lowest card being the seventh and the highest card being the ace. Each of the three active players receives 10 cards with the two remaining cards put aside as a talon that will be used by the declarer of a game allowing substituting for any of the two cards he had on hand before the game starts. When there are four players, in each hand, the dealer pauses. It is possible that only two players can play (*Gusarik* version in Russian), but still cards are dealt for three players. The third player is called the dummy; his cards are not opened during bidding. We are not going to discuss this particular type of play, but will concentrate on a normal three-person game.

After receiving their cards, the players start bidding to win the contract to play the game. The goal is to take the minimum of six tricks after declaring the trump suit. To organize a bidding process, it always starts with the eldest hand (whoever first got their cards). The ranking is first by number of tricks and then by suit as follows: spades, clubs, diamonds, hearts, and no trumps in ascending order. A special bid, *misère*, obligation to not take any tricks at all, ranks between eight tricks with no trumps and nine tricks with spades. If the bidding is going on between a player when he announces a *misère* and some other player wishes to play a regular game, that player can win the bid to play by announcing a nine-trick play. The bidding can last several rounds until all players but one decide to pass. The player who outbid the other players is called soloist. He declares the game to be played. His objective is to win the contracted number of tricks or *misère*, while the defenders' objective is to prevent this. According to the rules, the trick is won by the player who played the highest trump, or the highest card of the suit led. If a player does not have a card in a suit, there is an obligation to trump; if there is no trump, you can put down any card you wish. The eldest hand always leads to the first trick. The winner of a trick leads to the next trick. At the beginning of each round, the eldest hand can make a bid that only needs to be as high as the highest bid so far. Otherwise, each bid must be higher than the previous one. A player who has passed may not bid again later, and a player who wants to bid *misère* must not make any other bids before or after. In trick-play, the declarer must win at least the number of tricks indicated in the contract, 6, 7, 8, 9, or 10. If successful, the declarer wins the value of the contract in pool points 2, 4, 6, 8, or 10. If not successful, the declarer loses the value of the contract multiplied by the number of undertricks (tricks missing) in dump points, and also pays the same amount to each defender in whist points. Beginning with the player who sits to the left of the declarer, each defender indicates whether he or she wants to whist, i.e., is willing to play and win at least two tricks in a 6-trick game and one trick in a 7-trick game. Defenders, who declare whist, may drop out of trick-play if it is a 6- or 7-trick game. They also decide if they want to play with closed or open cards. A second incentive for whisting, besides the chance of spoiling the declarer's contract, is that the whisting players are paid the value of the contract in whist points from the declarer for each trick they win, regardless of whether the declarer or the defender wins their respective required number of tricks. If there is only one whister, then that player also gets the whist points for the tricks won by the other defender. In some variations of the game, there is convention that if the declarer lost the game the whists won by the whister are divided equally between the whister and the other player (assuming only one of the players declared whist). It is known as gentlemanly whist. The required number is four tricks if the declarer undertook to win six tricks, two tricks if the declarer undertook to win seven tricks, and one trick if the declarer's contract is for eight tricks or more. In case the defenders fail to win enough tricks, they are penalized for each undertrick. So, the whisting should be responsible.

The opposite of the normal game is the all-pass game (in Russian: *Raspasovka*). It is played when no player has made a bid. The objective is to win as few tricks as possible with no trumps. *Misère* and all-pass games are special in that the object is to avoid tricks rather than win them. In many cases, *misère* is considered as a very challenging game because of its value; it is regarded as having a contract value of 10, making it equivalent to winning a 10-trick game, which is a very rare game. However, a *misère* is considered to be lost even if one trick is taken (the value of each is negative 10 points) and it may become a very painful game for the loser if he takes several tricks. The defendants' goal is to "catch the *misère*", i.e., give as many tricks to the soloist as possible.

Let's now go over some specific game strategies and demonstrate how your plan for the game can get you to the desired goal or contrary, to a failure. It is all up to you how you decide to play and what combinations to consider getting for the result. The players should keep in mind the importance of the first move. Whoever is on the eldest hand always has an advantage of controlling the way the whole game is played. Sometimes, the first move becomes critical in a game's final outcome. What makes preferans particularly attractive is the fact that each player has to constantly pay attention to the cards that are still in play and the cards already played, causing the player to have to make immediate decisions regarding every next step depending on how the game goes. If some player makes a mistake, the other players have the chance to change the disposition and get a different result.

**Problem 4.** Since this is the first game-related problem, we suggest starting from the very beginning, meaning instead of formulating the problem to solve, let's analyze the disposition of cards on hands *before* any betting is started.

So, the original cards on hands before any bets are the following:

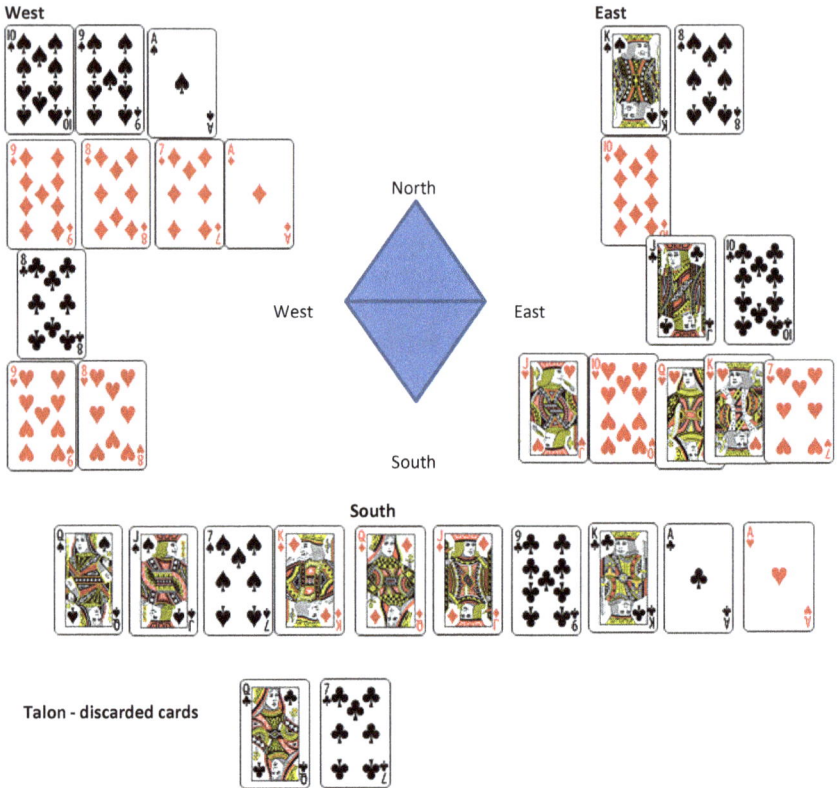

Clearly, West and East will pass; they have no good cards to declare any trump suit to get the minimum of six tricks. South has the last hand and finds himself in a dilemma whether to say "pass" or to play the game. His cards are much better, so he probably would declare a game and will take the talon, which happens to be a great addition for him. So now, he would discard two spades cards (he does not need them) and will declare the 7-trick game having clubs as a trump suit. One may even argue that he probably should declare an 8-trick game counting five tricks in clubs as in trump suit, the ace of hearts, and the king, queen, and jack of diamonds as two potential tricks, i.e., a total of eight tricks. Let's see if that would be justified in the given case.

So, we have now the following disposition:

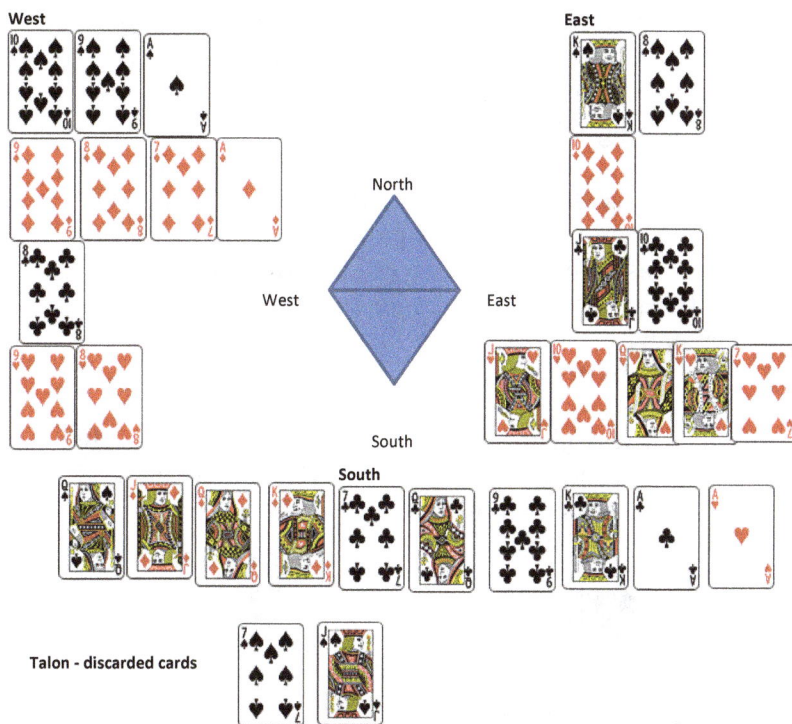

If South would have the eldest hand, he would win an 8-trick game. He goes four times with the ace, king, queen, and nine of clubs, his trumps. Then, he would go with the king of diamonds allowing the other players to take this trick for the ace of diamonds. The only other trick South would give in is the queen of spades. Since the other two players have no remaining clubs-trumps, South will take his last tricks with the last remaining trump, seven of clubs, ace of hearts, and the queen and jack of diamonds, completing his 8-trick game. However, if South is not on the eldest hand, then the whole game is different. Assume now that West goes first. He should go with the ace of diamonds. East would give up his ten of diamonds and South would give up his jack of diamonds. East now has no diamonds left, so the next move by West would be the nine of diamonds, and East will get the trick with his trump, the ten of clubs, for example; South would put down his queen of diamonds. This trick goes to East.

His next move has to be in the eight of spades. South has to hand in his queen of spades, and West will take this trick by putting down his ace of spades. West now goes with the eight of diamonds. East puts jack of clubs, his last trump, and South has to give his king of diamonds. This trick goes to East. His next move is irrelevant. All the remaining tricks will be South's because he has five clubs and the ace of hearts. The outcome of this game is pretty disastrous for South, because he collects only six tricks. What is interesting is that with the correct sequence of moves made by whisting West and East, South has no chance to collect more than six tricks. The same is true for the first scenario, assuming that South has the eldest hand. In that case, with the correct playing strategy, South should collect eight tricks no matter how West and East would play.

**Problem 5.** West plays a 6-trick game having clubs as the trump suit. South is on the eldest hand. Can West win assuming that the whisting players make no mistakes?

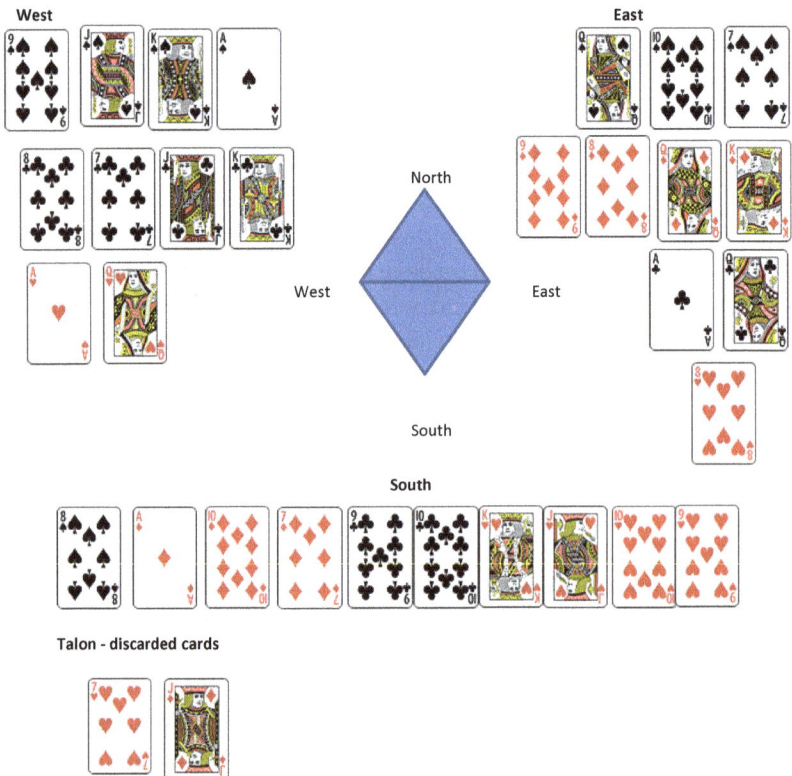

Talon - discarded cards

**Solution.** South goes with the ace of diamonds. West has to hit it with his trump, let's say, with the seven of clubs. East responds by putting down his eight of diamonds. Now, there are several playing options for West's next move. We leave the exercise to the readers to consider various scenarios. For example, West may go with the ace of hearts, taking this trick, and then go again with hearts, the queen of hearts, so East has to hit it with his queen of clubs as the trump. This is the first trick taken by the whisting players. East goes with the queen of diamonds next. West hit it again with his trump, the eight of clubs. South puts down his ten of diamonds. West now has only two trumps remaining. If he goes with the jack of clubs, East will hit it with the ace of clubs, and South will put down his nine of clubs. This will be the second trick taken by the whisting players. East should go next with the queen of diamonds, which West must hit with his last remaining trump, the king of clubs; South will put down his seven of diamonds. No matter how he plays next, West can take the only one trick for his ace of spades. The next move in spades will be hit by South's last trump, the ten of clubs, and he would take all the remaining tricks, because he will move to hearts. Since there are no more clubs-trumps remaining, all the last tricks will be his. So, in total, West would have only five tricks and will lose this game.

The winning strategy for whisting players is to stick to going with diamonds, thus reducing the trumps on hand of West. Then, regardless of his moves, West loses this game. However, if South would make his first move with a club, like the nine of clubs, then West gets a chance to win the game. No matter how the whisting players play after this first move, West is getting his six or even seven tricks by making the right plays. If with the second move East goes with the ace of clubs and the third move with the king of diamonds, then West would still have two clubs-trumps remaining. After taking the third diamonds trick, West should go with the ace of spades, take the fourth trick, and go with the nine of spades. East will hit it with his queen of spades and by going with one of his diamonds he would give that trick to West. Next, West collects his two tricks in spades, with the king and jack. In total, West gets his two tricks in clubs, a trick with the ace of hearts, and three tricks with spades, thereby winning the game. If in the second move East goes with diamonds, preserving the remaining trump, the ace of clubs, West would then need to go twice with hearts: first taking the trick with the ace of hearts and second with the queen of hearts, which East would take with his ace of clubs. In that scenario, West gives up only three tricks and wins seven.

**Problem 6.** West plays *misère*. East goes first. How should the game be played so that the bidding player gets tricks? How many tricks can be given to West assuming the goal is to give as many as possible?

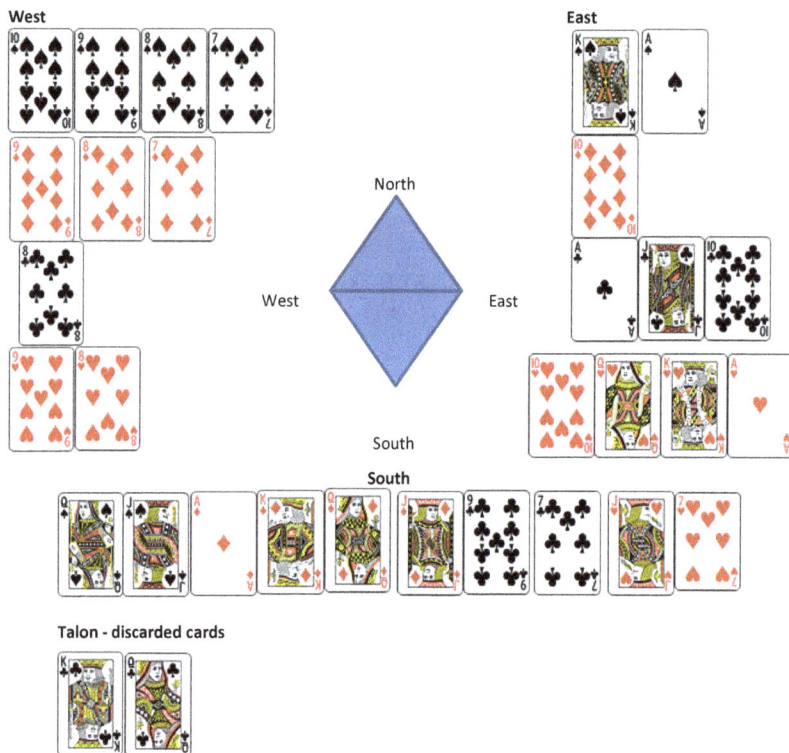

**Solution.** South goes three times with diamonds allowing East to discard his ace and king of spades on the second and third moves. So, now, East does not have any spades left. South makes the fourth move with the last remaining diamond and watches which card West will discard. If West throws down the eight of clubs, then East would throw out the ace of hearts. South's next move will be to go twice with the jack and queen of spades. West will put down his two spades cards, while East will throw out the king and queen of hearts. Next, South goes with the jack of hearts, collecting the nine of hearts from West and the ten of hearts from East. The final move will be to go with the seven of hearts, forcing West to take this trick with his eight of hearts. The remaining two tricks are also his. A total of three tricks are given in this case.

If on the fourth move, West discards one of his hearts cards, then East will throw out the ace of clubs. The next moves from South are the same; he goes twice with his spades cards, but now East will discard his two remaining club cards, the jack and ten. The next move is with South's jack of hearts taking eight of hearts from West and ten of hearts from East preserving the final move from South with the seven of clubs. West must take this trick with his eight of clubs, with no other choice as to get the final two tricks as well. In that scenario, again the total of his tricks becomes three!

For newcomers, the game might look complicated and confusing. It is true that one needs some practice to feel comfortable in analyzing and playing various combinations. The winning strategy is not always obvious from the first steps. It has to be noticed that it is easier to play with open cards, and it takes much more insight and understanding, sometimes even an intuition, to play the right way with the closed cards. There are some conventional "rules" for how to play in specific situations with closed cards. For example, while you are passing, and you are invited by the whisting player to play with closed cards against the soloist, being the eldest hand, you are supposed to go with your strong short suit first. This assumes that you would not help the soloist with his game, and you would most likely hit into his trumps. He then would need to make his next move by making guesses regarding the reasons behind the whisting player's decision to not open his cards (usually, one wants to play with closed cards if he has something to hide from the soloist; strong trump cards and lack of some suit, for instance). This may affect the playing strategy. Sometimes, an incorrect guess and the respective move can lead to one losing the whole game. By the way, since the passing game is always played with closed cards, it is considered more complicated than the normal game, and requires a great deal of attention and skill. *Misère*, as the most sophisticated of all games, is always played with open cards.

The beauty of preferans is that even when you follow logical agreed-upon steps, they will not necessarily guarantee the desired outcome. It all depends on the combinations of cards that are on hand, and the abilities of the players to properly analyze possible moves. The disposition might change with any move, and one has to immediately take advantage of open opportunities, even if it means going into a different direction than what was originally planned before. In many cases, your next move depends on the move you expect from another player (see Problem 6).

In my opinion, the amazing game of preferans is a great tool in developing logical thinking and problem-solving abilities. By way of analogy, when solving mathematical problems, one undertakes logical steps in a specific sequence to be able to adjust to the problem, and this can ultimately lead to finding an answer or a proof. If you focus on just making a single move at a time, achieving the end goal is almost impossible. On the other hand, by applying your knowledge and experience to see the bigger picture, you can come up with various strategies to bring you to the end goal.

Unfortunately, I did not find any literature in English related to preferans, but there are several software versions of the game, providing various levels of difficulty. Intellectual game lovers should enjoy this activity.

A few problems — compositions are offered below for readers to analyze and find the right strategy for the optimal outcome.

**Problem 7.** West plays a 9-trick game having spades as the trump suit. East is on the eldest hand. How many tricks should East and South allow for West assuming the right playing strategy?

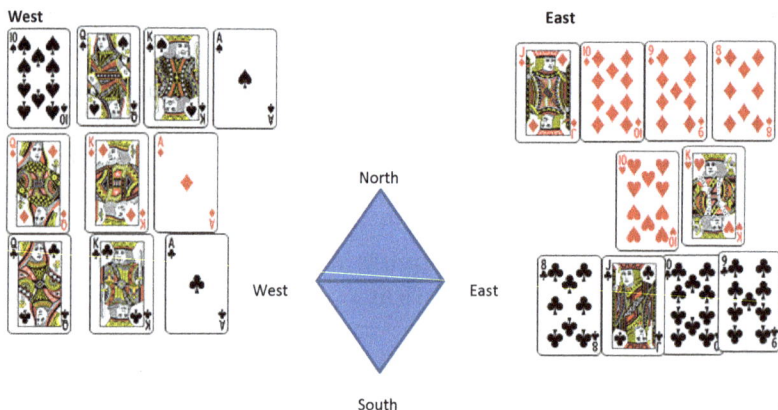

**South**

Talon - discarded cards

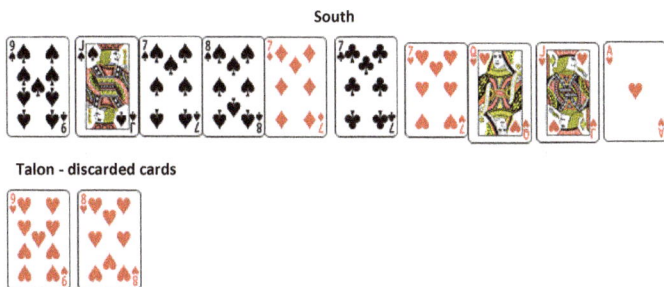

**Problem 8.** West plays an 8-trick game having spades as the trump suit. South is on the eldest hand. Can West win this game?

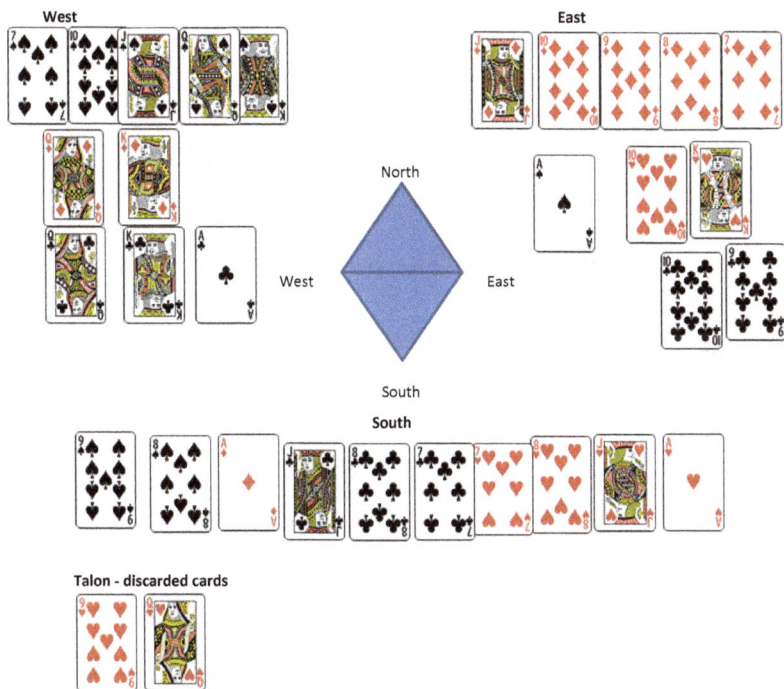

**Problem 9.** It was a betting war between West and East to decide who is going to play the game. Finally, East outbid West and announced an 8-trick game having diamonds as the trump suit. Can East win this game, assuming that West is on the eldest hand? Would it matter if East would be on the eldest hand? Do you agree with East's decision to play an 8-trick game, not a 7-trick game?

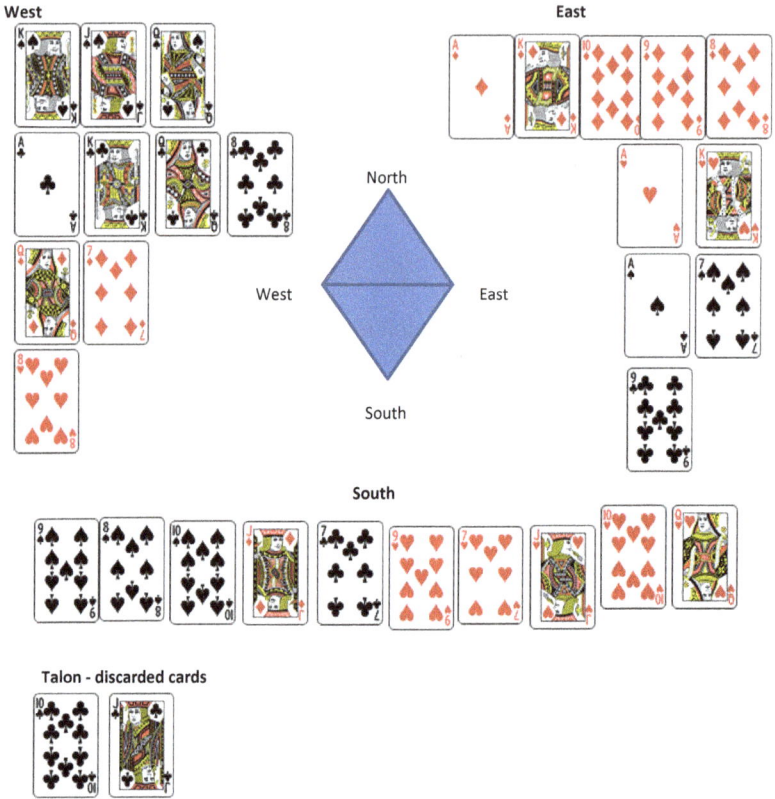

## Chapter 6

# Euclidean Plane Transformations

In his brilliant book *How To Solve It*, George Polya (1887–1985), while speaking about the benefits of studying geometry and its applications to logical system, said that "... the system of geometry is cemented with proofs. Each proposition is linked to the foregoing axioms, definitions, and propositions by a proof. Without understanding such proofs we cannot understand the very essence of the system. In short, if general education intends to bestow on the student the idea of logical system, it must reserve a place for geometric proofs". I cannot agree more. Studying geometry provides the great opportunity for developing logical thinking and practicing mental operations that build rigorous proofs.

This chapter will be devoted to practical applications in problem-solving of isometric and homothetic transformations properties. While Euclidean plane transformations are briefly presented in high school curriculum, their practical applications in problem-solving are almost completely omitted. I find this very unfortunate because studying the applications of Euclidean plane transformations offers a rewarding and very useful alternative to conventional methods and techniques for many interesting geometrical problems.

Before we proceed, it worth emphasizing that in solving all of the construction problems in our book we assume that readers are familiar with basic constructions performed with the straightedge and compass. Let's just briefly refresh here the meaning of Euclidean constructions. Any straight line is defined by two distinct points and can be constructed with the use of a straightedge (unmarked ruler). A circle is determined by its center and the length of its radius and can be drawn with the use of a compass (more correct, "pair of compasses"). We can measure the length of a segment only with a compass by preserving a radius length. All of our construction steps have to be justified; no approximations are acceptable. During our solutions, we will refer to some simple constructions assuming readers are well acquainted with them from high school curriculum. We are not going to explain each of those constructions in greater detail, but strongly recommend that readers thoroughly perform every step in the construction process and its justification, determine how unique the solution is, and review alternative solutions. Instead, we will concentrate on problems' analyses and strategies, explaining in detail every decision to be made and every step to follow.

## 6.1 Isometric Transformations

By isometric plane transformation, we understand a distance-preserving transformation in the plane. Reflections, rotations, translations, and combinations of them are the types of Euclidean plane isometric transformations.

A *reflection through an axis* transforms any point on a plane into its image such that the distance from each to the axis, line of symmetry, is the same.

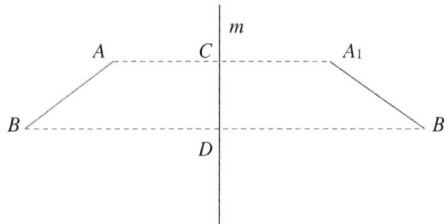

In the above figure, the image of segment $AB$ in reflection through $m$ axis is the segment $A_1B_1$ such that $AC = CA_1$, where $AC \perp m$, $CA_1 \perp m$, $(C \in m)$ and $BD = DB_1$, where $BD \perp m$, $DB_1 \perp m$, $(D \in m)$. In this reflection, $A_1$ is an image of $A$, $B_1$ is an image of $B$, and, respectively, $A_1B_1$ is an image of $AB$. Since any isometric transformation preserves the distance between the corresponding points, $A_1B_1 = AB$.

By the definition of *rotation*, the entire plane is turned about some point, the center of rotation, through an angle clockwise or counterclockwise. The size and shape of any figure are kept invariant, but its points all move along arcs of concentric circles. The center, which may or may not belong to the figure being rotated, is the only point that remains fixed.

Let's demonstrate a simple rotation of a segment $AB$ about a point $O$, the center, through an angle $\alpha$ (see the following figure). Point $A$ is rotated into point $A_1$ and $B$ into $B_1$ such that $OA = OA_1$ and $\angle AOA_1 = \alpha$; $OB = OB_1$ and $\angle BOB_1 = \alpha$. $AB$ is transformed into $A_1 B_1$.

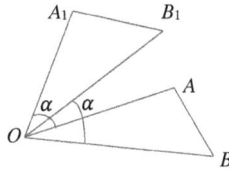

Because rotation preserves distance, it takes any figure into a congruent figure. Therefore, $AB = A_1 B_1$ and consequently, triangle $AOB$ is congruent to triangle $A_1 O B_1$. The angles between corresponding lines are equal to each other and to the given angle of rotation. Thus, an angle between $AB$ and $A_1 B_1$ is equal to $\alpha$ as well.

Out of all rotations, we distinguish the rotation that transforms every line into a line parallel to it. This is the *half-turn*, or rotation through 180°, which transforms each ray into an oppositely directed ray. Since a half-turn is completely determined by its center, another name for the half-turn is *central symmetry*. A point and its image are called symmetric with respect to the center of rotation.

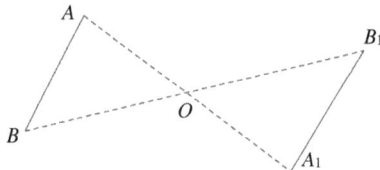

In the above figure, $AO = OA_1, BO = OB_1$. Points $A_1$ and $B_1$ are the images of points $A$ and $B$ in a central symmetry with the center $O$. Clearly, $AB = A_1 B_1$ and $AB \parallel A_1 B_1$.

Finally, a *translation* is a geometric transformation that moves every point of a figure by the same distance in a given direction.

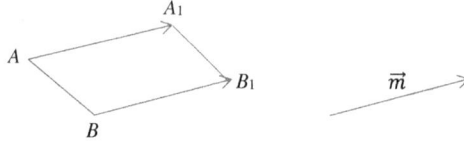

Translation as a geometric transformation is associated with a Euclidean vector, an entity defined by its length and a direction. In the above figure, vector $\vec{m}$ endows the direction and the magnitude of the specific translation to be applied to segment $AB$. In order to obtain images of $A$ and $B$, the points $A_1$ and $B_1$, we need to move points $A$ and $B$ in the given direction, and the length of each segment $AA_1$ and $BB_1$ has to equal the length of the given vector $\vec{m}$. Being an isometric transformation, translation preserves the distance between the respective segments; therefore, $AB = A_1B_1$. By the definition of translation, all respective segments are parallel. Translation preserves the shape and size of any figure.

All the problems considered below have multiple solutions. We will illustrate the applications of isometric transformations and explain the reasoning and usefulness of each such application. We strongly recommend that readers search for alternative solutions, and analyze and compare their findings with the suggested techniques.

**Problem 1.** There is given a straight line $a$ intersecting the given segment $AB$ at $M$. Given that the bisector of an angle $C$ is contained on $a$, construct triangle $ABC$.

**Solution.** This construction calls for determining the triangle when one side, $AB$, is given as well as the line possessing the angle bisector of angle $C$.

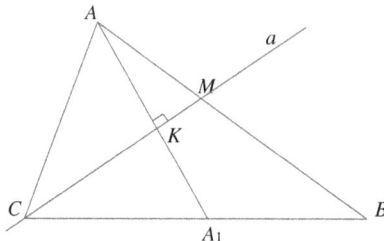

To solve the problem, we will apply the well-known property of an angle bisector in a triangle: *a point lies on an angle bisector if and only if it is equidistant from its sides.* Constructing point $A_1$, the image of point $A$ in reflection through $a$ axis, we know that it must be located on $BC$. Drawing a line connecting $B$ and $A_1$ till it intersects with line $a$, we get the third vertex $C$ of the desired triangle $ABC$.

The construction steps are as follows:

Draw $AK \perp a$ and locate $A_1$ on extension of $AK$ such that $AK = A_1K$ ($K$ lies on line $a$).
Draw $BA_1$ till its intersection with $a$ at $C$. Finally, draw $CA$.

Let's prove that $ABC$ is the desired triangle. Indeed, we obtained two congruent right triangles $AKC$ and $A_1KC$ because $CK$ is the common leg and $AK = A_1K$ by construction. Therefore, $\angle ACK = \angle A_1CK$, which means that $CK$ lies on the angle bisector $CM$, and triangle $ABC$ satisfies all the given conditions. Now, we have to determine how many solutions the problem may have depending on the location of $a$ and $AB$. If $a$ is not perpendicular to $AB$, as it is shown in the figure above, then only one triangle can be constructed. Should $a$ be perpendicular to $AB$, then $M$ will necessarily be the midpoint of $AB$. This is because it is given that the angle bisector of angle $C$ lies on $a$, so if $AB \perp a$, then $ABC$ has to be an isosceles triangle having $CM$ as its angle bisector, altitude, and median at the same time. Any point on $a$ can be selected as the third vertex of the triangle, and we get infinitely many triangles, all of which are isosceles triangles satisfying the given conditions.

**Problem 2.** There are two balls $A$ and $B$ located on a pool table. In what direction should you hit ball $A$ with the cue stick so it hits ball $B$ after it bounces off a selected edge of the pool table? Assume that when the ball is directed at the pool table's edge in an angle $\alpha$, it bounces off this edge at the same angle $\alpha$ as well (see the following figure).

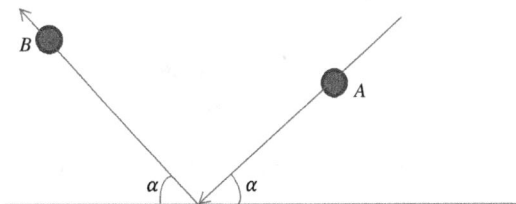

**Solution.** Our goal is to send the ball from point $A$ to the pool table's side $MN$ in such a way that after hitting this side it will be bounced into a ball placed at a specific point $B$ on the table.

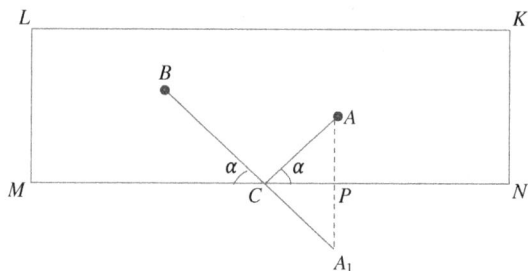

In geometrical terms, the problem is about locating point $C$ on the side of a rectangle $MLKN$ such that angles $MCB$ and $NCA$ are congruent. We have to find $A_1$, the image of $A$ in reflection through the $MN$ axis. Draw $AA_1 \perp MN$. Let $P$ be the point of intersection of $MN$ and $AA_1$. We locate $A_1$ such that $AP = PA_1$. Now, draw a straight line through $A_1$ and $B$ and denote $C$ the point of intersection of $A_1B$ with $MN$. By construction, in triangles $APC$ and $A_1PC$, we have $\angle APC = \angle A_1PC = 90°$, $AP = A_1P$, and $PC$ is the common leg. Therefore, right triangles $APC$ and $A_1PC$ are congruent. Hence, they have respective equal angles $ACP$ and $A_1CP$, $\angle ACP = \angle A_1CP = \alpha$. Connecting $B$ and $A_1$, we found $C$ as the point of intersection of $BA_1$ and $MN$, such that $\angle MCB = \angle A_1CP = \alpha$ as vertical angles. Therefore, due to transitivity, we see that $\angle MCB = \angle ACP = \alpha$, and we arrive at the conclusion that $C$ is the desired point.

Why did we use reflection to solve the problem, and how did we decide it would be helpful? Well, as in many problems involving rectangles and squares, reflection might help to identify some auxiliary elements useful in linking together the given conditions that are not seen before. In this case, it was an auxiliary angle $A_1CP$, which immediately clarified the whole picture. Furthermore, applying a reflection allows us to draw another important conclusion that the problem will have a solution only when $B$ is not located on the perpendicular dropped from $A$ to $MN$.

**Problem 3.** Given the same conditions as in Problem 2, but now the question is in what direction should you hit ball $A$ with a cue stick so it hits ball $B$ after bouncing off the two selected parallel edges of the pool table?

**Solution.** Our goal is to find point $C$ on $MN$ and $D$ on $LK$ such that after hitting sides $MN$ and $LK$ at $C$ and $D$, respectively, the ball directed from $A$ will hit the ball located at $B$. Our analysis in Problem 2 will be useful in this case as well. However, now we need to consider two reflections.

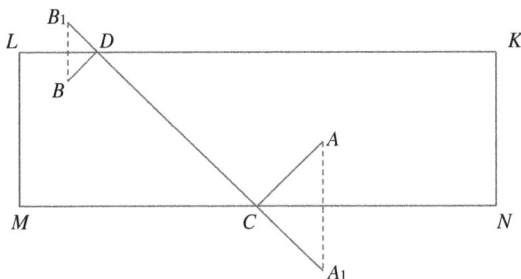

First, we find $A_1$, the image of $A$ in reflection through $MN$. Secondly, we find $B_1$, the image of $B$ in reflection through $LK$. Next, by drawing $A_1B_1$, we locate its points of intersection with $MN$ at $C$ and with $LK$ at $D$. $C$ and $D$ are the desired points. Indeed, it is easy to prove, and we leave that exercise for the readers, that $\angle A_1CN = \angle ACN = \angle BDL = \angle B_1DL$. This proves that if we directed the ball from $A$ to $B$ as explained above, its trajectory will consist of segments $AC$, $CD$, and $DB$. The problem has no solutions when $AB$ is perpendicular to the indicated parallel sides of the pool table $MN$ and $LK$; otherwise, the solution is unique.

How many of our pool game fans realize that by applying the discussed techniques, they would achieve the best possible results for their game? Actually, whether you like it or not, while setting a specific trajectory for each hit, you are "inadvertently" applying reflection properties. The more precise you are in utilizing these properties, the better your result will be!

Applying rotations and central symmetry usually proves useful in problems involving regular polygons or figures with equal angles.

In comparison with conventional methods (we recommend the readers to investigate alternative solution(s) of the problems), rotations allow for relatively simple and elegant proofs based on the rotation properties of preserving the distances and angles between corresponding linear elements.

**Problem 4.** Find angle $C$ in a triangle $ABC$, being given $\angle A = 60°$ and $AC = \frac{1}{2}AB$.

**Solution.** The half-turn about $C$ as the center takes $A$ into $D$ such that $AC = CD$ and all three points lie on one line. It is given that $AC = \frac{1}{2}AB$. Therefore, $AD = AB$. In isosceles triangle $BAD$, $\angle A = 60°$, hence it is an equilateral triangle. By construction, $C$ is the midpoint of $AD$. It implies that $BC$ is the median in the equilateral triangle $ABD$. By the properties of the equilateral triangle, any of the triangle's medians is an angle bisector and an altitude all at the same time. Therefore, $BC \perp AD$ and we conclude that $\angle ACB = 90°$.

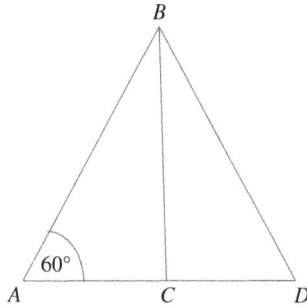

Why did we use the rotation about $C$ through $180°$? How did we come up with that decision? The answers to these questions are hidden in the problem's conditions. It was given that $AC = \frac{1}{2}AB$, so finding $D$ as the image of $A$ in such a rotation allowed us to introduce an auxiliary triangle $ABD$, one of the angles of which was given to equal $60°$. By performing the half-turn, we arrived at not just an isosceles but an equilateral triangle $ABD$. In the final steps in our solution, we utilized the properties of an equilateral triangle to arrive at the desired result for $\angle ACB = 90°$.

**Problem 5.** Given an equilateral triangle $ABC$ and an interior point $M$ such that $\angle AMB = 120°$, find the length of $CM$ knowing that $AM = 1$ and $BM = 2$.

**Solution.**

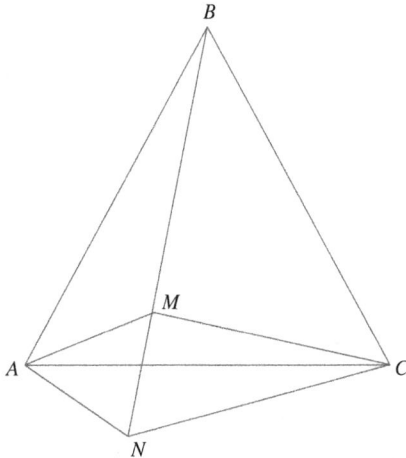

This is a more complicated problem than the previous one. At first glance, it's not clear how can we relate the given elements to find $CM$. As in many geometrical problems, when you are not sure where to start, it should be helpful to introduce some auxiliary elements, and then do some additional constructions which would allow for identifying the links and connections not seen before. In some problems with given angles of $30°, 60°, 90°$, and $120°$, it's not a bad idea to try applying rotations. This will create new triangles, squares, quadrilaterals, and other figures that could be helpful in revealing bridges between the given elements, and may clarify the next steps to take in a solution process.

It is given that $ABC$ is an equilateral triangle, so all of its angles equal $60°$. A $60°$ clockwise rotation about $A$ takes $B$ into $C$, and $M$ into $N$ such that $AM = AN$ and $\angle MAN = 60°$. The obtained triangle $AMN$ is isosceles with a $60°$ angle between the equal sides $AM$ and $AN$. Therefore, $AMN$ is an equilateral triangle and $MN = AM = AN = 1$. Since our rotation takes $B$ into $C$ and $M$ into $N$, then $BM$ is rotated into $CN$, $AB$ into $AC$, $AM$ into $AN$, and we conclude that triangle $AMB$ is rotated into $ANC$. Rotation preserves the distances and angles between the corresponding rotated segments. Therefore, $CN = BM = 2$, $AM = AN = 1$, $\angle ANC = \angle AMB = 120°$.

Since $\angle ANC = \angle ANM + \angle MNC$, we get that

$$\angle MNC = \angle ANC - \angle ANM = 120° - 60° = 60°.$$

So, in triangle $NMC$, we have $MN = 1$, $CN = 2$, and $\angle MNC = 60°$. We got a triangle with two sides forming an angle of $60°$ and one of these sides is twice as big as the other side. According to the previous problem, $\angle NMC = 90°$. Applying the Pythagorean Theorem to the right triangle $NMC$, we arrive at $CM = \sqrt{2^2 - 1^2} = \sqrt{3}$.

**Problem 6.** From the vertex $A$ of a square $ABCD$, two rays are drawn inside the square. From vertices $B$ and $D$, perpendiculars are dropped to the two rays: $BK$ and $DM$ are dropped to one of them, and $BL$ and $DN$ are dropped to the other. Prove that the segments $KL$ and $MN$ are congruent and perpendicular.

This problem was offered by D. Nyamsuren in the currently defunct magazine *Quantum*, March/April 1994 issue.

**Solution.**

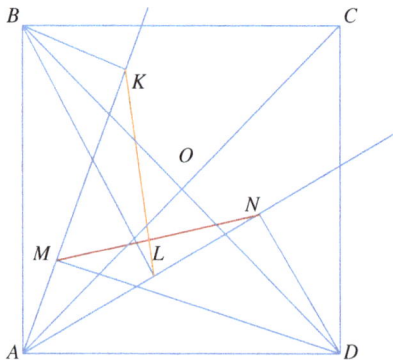

Analyzing the problem's data, we see that there are many perpendicular segments and right angles involved. This should be a hint for us to try to apply rotation to simplify the solution. If you manage to select the center of rotation in such a way that the segments in question will be rotated into each other in a rotation by a $90°$ angle, then they will satisfy the conditions in the question asked, and the problem will be solved.

Considering a square, the best results are usually achieved for a 90° rotation about its center $O$, the point of intersection of its diagonals. This will hold true in our problem as well. Consider the counterclockwise 90° rotation of $ABCD$ through $O$ and show that it will take the segment $MN$ into $KL$.

In the right triangle $AKB$, $\angle KBA = 90° - \angle KAB$. On the other hand, since $\angle BAD = 90°$, then $\angle KAD = 90° - \angle KAB$ (angle $KAD$ is supplemented by the angle $KAB$ to 90°) and we see that $\angle KBA = \angle KAD$. It follows that the right triangles $BKA$ and $AMD$ are congruent by the side $(AB = AD)$ and two adjacent angles (clearly, $\angle MDA = \angle KAB$ as well), and in our rotation $\triangle BKA$ is rotated into $\triangle AMD$. Indeed, $B$ will be rotated into $A$ and $A$ into $D$. The ray $BK$ is rotated into the ray $AM$ because $BK \perp AM$ by the given conditions (in a 90° rotation taking $B$ into $A$, the ray $AM$ is taken into the perpendicular ray $BK$). Likewise, since $A$ is rotated into $D$ and $AK \perp DM$, then $AK$ is rotated into $DM$. Therefore, the intersection point $K$ of $BK$ and $AM$ is rotated into the intersection point $M$ of $AK$ and $DM$. In the same way, we can prove that $L$ is rotated into $N$. Hence, we obtained that the segment $KL$ is rotated into the segment $MN$. Since rotation preserves distances and angles between the corresponding rotated segments, we arrive at the desired result, $MN = KL$ and $MN \perp KL$.

Next, we will turn to the properties of translations and demonstrate their applications in problem-solving.

**Problem 7.** Given two non-intersecting circles located outside of each other, a segment, and a straight line, non-parallel to the given segment, construct a segment parallel to the given line such that it has the same length as the given segment and its end points are located on each of the given circles.

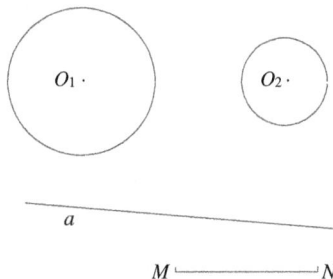

**Solution.** This construction calls for drawing a segment of the length equal to $MN$ parallel to line $a$, such that its end points are located on the circles with centers $O_1$ and $O_2$ as shown in the above figure.

The solution is greatly simplified by applying the properties of translations. A translation transformation involves moving the geometrical objects in the same direction determined by the specific vector. All respective straight lines are parallel, so when facing parallelograms or trapezoids (given some parallel segments or parallel straight lines), it makes sense to attempt to utilize translation properties.

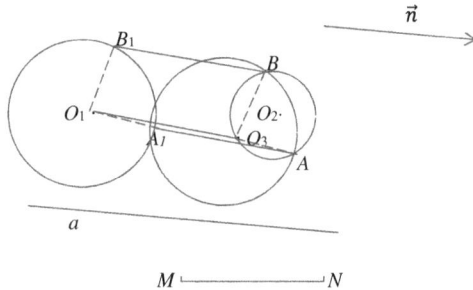

Consider a vector $\vec{n}$ such that $\vec{n} \parallel a$ and $\|\vec{n}\| = MN$. Construct a circle with the center $O_3$, the image of the circle with the center $O_1$ in a translation by vector $\vec{n}$. Denote by $A$ and $B$ the points of intersection of that circle with the circle with the center $O_2$. Draw two straight lines through $A$ and $B$ parallel to $a$, and locate on them points $A_1$ and $B_1$ such that $AA_1 = BB_1 = MN$. Let's prove that $A_1$ and $B_1$ are the points of intersection of these straight lines with the circle with the center $O_1$. By doing this, we will prove that segments $AA_1$ and $BB_1$ are the desired segments and the problem will be solved.

In the first step, we constructed $O_1O_3 \parallel \vec{n}$ and $O_1O_3 = \|\vec{n}\| = MN$. After finding points of intersection $A$ and $B$ of the circles with the centers $O_3$ and $O_2$, we constructed $AA_1 \parallel BB_1 \parallel MN$ and $AA_1 = BB_1 = MN$. Therefore, $O_1O_3AA_1$ and $O_1O_3BB_1$ are the parallelograms because the pairs of their opposite sides are parallel and have equal lengths. So, the other pairs of opposite sides must be parallel and of equal lengths as well, $O_1A_1 = O_3A$ and $O_1B_1 = O_3B$. Recall now that $A$ and $B$ were obtained as the points of intersection of the circles with the centers $O_3$ and $O_2$. Since the circle with the center $O_3$ was an image of the circle with the center $O_1$ in the translation by vector $\vec{n}$, their radii are of the same length, which means the lengths of each segment $O_1 A_1$ and $O_1 B_1$ are equal to the radius of the circle with the center $O_1$. In other words, we just proved that $A_1$ and $B_1$ are the points located on the circle with the center $O_1$, which was to be proved. So, when the image of the first circle in translation by the given vector intersects the other given circle in two points, the problem has two solutions.

Finally, it has to be observed that if the circles with the centers $O_3$ and $O_2$ are tangent, i.e., have only one point of intersection, the problem will have one solution; we will be able to construct only one segment of the given length parallel to the given line and with end points located on the two given circles. If the circles do not intersect, the problem has no solutions.

**Problem 8.** A convex quadrilateral is cut along a diagonal; a congruent quadrilateral is cut along the other diagonal. Put the four pieces together to make a parallelogram.

This problem was offered by V. Proizvolov in the currently defunct magazine *Quantum*, September/October 1992 issue.

**Solution.**

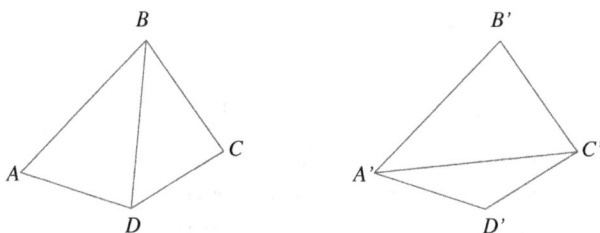

It is given that $ABCD$ and $A'B'C'D'$ are congruent convex quadrilaterals. Our goal is to make a parallelogram from four triangles $ABD$, $BCD$, $A'D'C'$, and $A'B'C'$ (see the above figure).

Translating triangle $ABD$ by vector $\overrightarrow{AC}$ into triangle $CB_1 D_1$, we deliver the sought-after parallelogram $BB_1 D_1 D$.

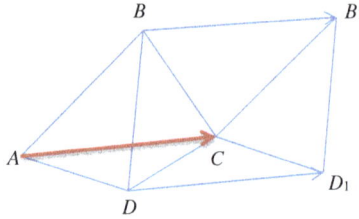

Indeed, by the properties of the translations, $BD \parallel B_1D_1$ and $BB_1 \parallel DD_1$, so $BB_1D_1D$ is a parallelogram by definition. Its parts $BCD$ and $B_1CD_1$ (congruent to $BAD$) are taken from the quadrilateral $ABCD$, and $BCB_1$ (congruent to $A'B'C'$) and $DCD_1$ (congruent to $A'D'C'$) are taken from the second quadrilateral $A'B'C'D'$.

Before we continue our discussion of practical applications of isometric transformations to homothetic transformations, it merits mentioning that various compositions of isometric transformations may be very useful in many geometrical problems as well. One such example, called *glide reflection*, is a composition of translation with a reflection over a line. It can be considered both ways, translation followed by a reflection over a line, or vice versa. The order is not important. This composition is isometry, so it preserves distances. It also maps parallel lines onto parallel lines and preserves angles.

The reader may wish to explore glide reflection as well as the considered other isometric transformation applications and similar transformations further.

## 6.2   Homothety

Homothety or dilation is a transformation of Euclidean plane with respect to some fixed point $O$, the center of homothety, and a ratio $k \neq 0$ which brings each point $A$ into one-to-one correspondence to its image point $A'$ on the straight line $OA$ such that

$$OA' = k \cdot OA.$$

If $|k| > 1$, this transformation is called an expansion,
If $|k| = 1$, it is an identity transformation,
If $|k| < 1$, this transformation is called a contraction.

Contrary to isometric transformations, homothety may alter distances. All distances are increased or decreased in the same ratio $k$, when $k \neq 1$; it is called the ratio of magnification or dilation factor. Homothetic transformations preserve angles and orientation, and change each figure into a similar figure.

A homothety maps a straight line passing through its center into itself and a straight line not passing through its center into a straight line parallel to it. Under a homothety, segments are mapped into parallel segments with a length which is $|k|$ times the original length.

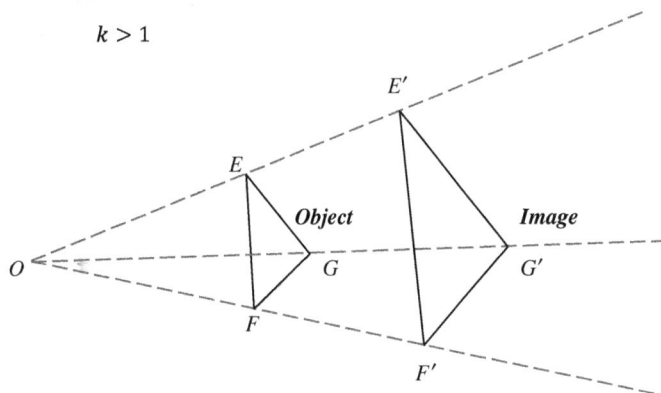

Considering $O$ as the center of a homothety with the coefficient $k > 1$ in the above figure, we obtain the triangle $E'F'G'$, the homothetic image of the triangle $EFG$, such that all the corresponding sides of both triangles are parallel, and each side of $E'F'G'$ is $|k|$ times the length of the respective side of $EFG$. Triangles $EFG$ and $E'F'G'$ are similar with the coefficient of proportionality $k$.

Homothetic transformation is useful for proving collinearity, concurrency, and determining ratios. It is also helpful in simplifying the solutions of many construction problems involving similar figures.

We will start with a problem about the property of a straight line passing through the points of intersection of the diagonals of a trapezoid and the point of intersection of its non-parallel sides. It has a short and elegant solution derived from homothetic properties.

**Problem 9.** (Jacob Steiner's (1796–1863) problem) Given a trapezoid $ABCD$ ($AD \parallel BC$), $P$ is the point of intersection of the non-parallel sides $AB$ and $DC$, and $O$ is the point of intersection of diagonals $AC$ and $BD$. Prove that the straight line $PO$ divides $AD$ and $BC$ into two equal parts.

**Solution.** Let's consider two homothetic transformations, homothety with the center $P$ and homothety with the center $O$. In the first homothety, the image of $BC$ will be $AD$. To prove this, we merely have to recall the definition of a homothety and to observe that $AD \parallel BC$ and $B$ and $C$ lay on lines $PA$ and $PD$, respectively. In a similar fashion, the homothety with the center $O$ will map $BC$ into $AD$; in this case, the coefficient will be a negative number, $k < 0$. Obviously, the midpoint of a segment will be brought into the midpoint of its homothetic image because under a homothety all the distances between points are multiplied by the same number $|k|$.

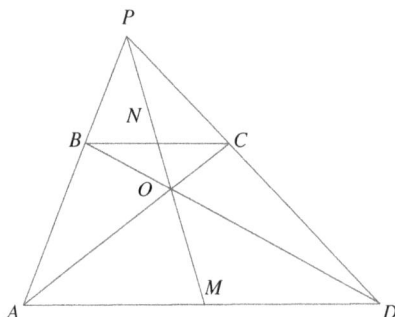

Finally, the midpoints of the bases of the trapezoid, being the respective points in a homothety, lay on the straight line passing through its center; in the first homothety, it is the straight line passing through $P$, $N$, and $M$, and in the second homothety, it is the straight line passing through $O$, $N$, and $M$. Since two points $M$ and $N$ define a unique straight line, then all four points $P$, $N$, $O$, and $M$ are on the same line. This concludes our proof.

Relying on the obtained result, let's examine homothety applications in two more classic problems, providing the alternative to conventional methods for proofs of the concurrence of the medians and the altitudes of a triangle.

**Problem 10.** Prove that the three medians in a triangle meet in a point called the centroid of the triangle, which is $\frac{2}{3}$ of the distance from any vertex to the midpoint of the opposite side. That is, the medians of a triangle divide each other in the ratio 2:1.

**Solution.** It is given that $AD$, $BE$, and $CF$ are the medians (segments joining each vertex to the midpoint of the opposite side) of triangle $ABC$. Our goal is to prove that they are concurrent and the point of their intersection splits each median in the ratio 2:1.

By the properties of a midline of a triangle (the segment that connects two midpoints of two sides), it is parallel to the opposite side and is half as long.

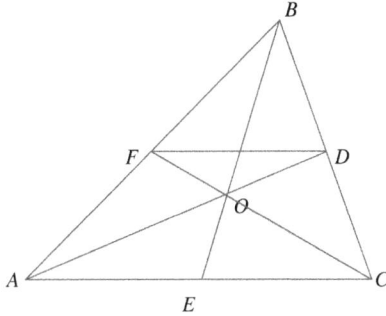

So, $AFDC$ is a trapezoid, in which $FD \parallel AC$ and $FD = \frac{1}{2}AC$. Its diagonals $AD$ and $CF$ intersect at point $O$. Non-parallel sides of $AFDC$ intersect at $B$. As was proved in the previous problem, $O$ lies on the line passing through $B$ and the midpoint $E$ of $AC$. Therefore, $O$, the point of intersection of the medians $AD$ and $CF$, belongs to the third median $BE$ as well. This completes the proof of the concurrency of three medians of a triangle. Since $FD \parallel AC$ and $FD = \frac{1}{2}AC$, the homothety with the center $O$ and coefficient $k = -\frac{1}{2}$ transfers $AC$ into $DF$. Therefore, $\frac{DO}{OA} = \frac{FO}{OC} = \frac{EO}{OB} = \frac{1}{2}$, and we arrive at the desired result.

**Problem 11.** Prove that the three altitudes in a triangle are concurrent at a point called the orthocenter of a triangle.

**Solution.** Consider triangle $ABC$. Let's draw its medians $AD$, $BE$, and $CF$ and its altitudes (perpendiculars dropped from the vertex to the opposite side) $BB_1$ and $CC_1$.

Denote by $O$ the centroid of $ABC$. In a homothety with the center $O$ and the coefficient $k = -2$, the midpoints $F$, $D$, and $E$ are brought into the vertices of the triangle $ABC$, $C$, $A$, and $B$; the altitudes of the triangle $FDE$ are brought into the altitudes of the triangle $ABC$. Let's draw $EE_1 \perp FD$ and $FF_1 \perp ED$. These altitudes of the medial triangle $EFD$ at the same time are the perpendicular bisectors of the original triangle $ABC$ because each passes through the midpoint of the respective side and is perpendicular to the opposite midline, which is the parallel to that side.

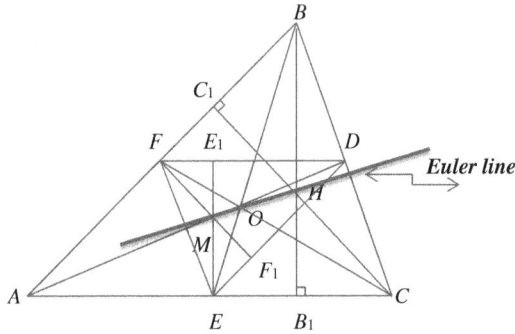

We know that three perpendicular bisectors to a triangle's sides are concurrent and the point of their intersection is the center of a circumscribed circle over a triangle. Therefore, the image of $M$ in the considered homothety will be the point of intersection of the altitudes of $ABC$, point $H$, and we conclude the proof of the orthocenter's existence.

Amazingly, we just proved in passing the fact of the existence of the famous *Euler line*, that the centroid, orthocenter, and the circumcenter of a triangle are collinear; the centroid lies between the circumcenter and the orthocenter and is twice as far from the orthocenter as it is from the circumcenter, $HO = 2OM$.

In this chapter, we restricted our examination of the isometric and homothetic transformations to those that are related to the two-dimensional Euclidean plane. Ambitious readers can extend this study to the three-dimensional Euclidean space and explore the analogies of the considered properties in solid geometry.

To acquire practice in the studied applications of the plane transformations, we offer several problems below for the readers to try out. Even though each of these problems has multiple solutions, which we encourage the readers to find, keep in mind that all of them can be solved utilizing the techniques examined in this chapter (we provide specifically those examples at the end of the book in the "Solutions to Problems" section).

**Practice exercises.**

**Problem 12.** Given three straight lines $a$, $b$, and $c$ intersecting each other, construct a segment perpendicular to $b$ such that its end points are located on $a$ and $c$ and its midpoint is located on $b$.

**Problem 13.** There are three given points $A$, $B$, and $C$ located on the same straight line ($B$ is between $A$ and $C$). Two equilateral triangles $ABC'$ and $BCA'$ are built above the straight line passing through $A$, $B$, and $C$. $M$ is the midpoint of $AA'$ and $N$ is the midpoint of $CC'$. Prove that $\triangle BMN$ is an equilateral triangle.

**Problem 14.** In a convex quadrilateral $ABCD$, points $E$ and $F$ are arbitrarily selected on the sides $AB$ and $CD$, respectively. Prove that the midpoints of segments $BF$, $AF$, $CE$, and $DE$ are the vertices of a convex quadrilateral whose area does not depend on a particular selection of $E$ and $F$ on $AB$ and $CD$.

**Problem 15.** There are given points $O_1$, $O_2$, $O_3$, and $A$. Construct $A_1$, the image of $A$ in point symmetry with respect to the center $O_1$, then $A_2$, the image of $A_1$ in point symmetry with respect to the center $O_2$, then $A_3$, the image of $A_2$ in point symmetry with respect to the center $O_3$. Repeat similar constructions for $A_3$, that is, find its consecutive images in point symmetries with respect to each center $O_1$, $O_2$, and $O_3$. Prove that the last such image will coincide with $A$.

**Problem 16.** Given a straight line and two points outside of it, locate a point on this straight line such that the sum of distances from it to the two given points is minimal.

**Problem 17.** Construct a triangle given its three medians.

**Problem 18.** (Offered by V. Proizvolov in *Quantum*, November/December 1994).

Two isosceles right triangles are brought together as shown in the figure below. Prove that the midpoints of the sides of the non-convex quadrilateral they form are the vertices of a square.

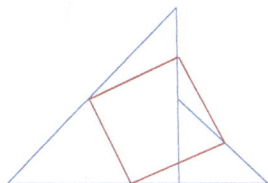

## Chapter 7

# Geometrical Constructions with Restricted Elements

The most intriguing and challenging problems in geometry in my opinion relate to so-called Euclidean constructions — constructions that are performed with two classic tools of a straightedge and compass. This topic is not just one of the soundest forms of problem-solving in geometry, but it also is one of the most illustrative in terms of coming up with solving planning strategies and analysis.

In this chapter, we will provide analyses of several non-trivial construction problems in which there are inaccessible elements. These problems stand out from the other construction problems because of the inability to use some restricted elements. For example, when it is stated in a problem that the vertex of a polygon is inaccessible, we can't use that point in any of our constructions, even though it seems like we can clearly see its position in a picture. We can surely utilize any other points on a polygon's sides, except for its vertex. In all of our constructions, we will use only a compass and straightedge unless it is specifically indicated otherwise. As in the previous chapter, the explanations of basic constructions will be omitted. However, we recommend that readers thoroughly perform every step in a construction process and its justification.

**Problem 1.** Given a parallelogram with four inaccessible vertices, find the point of intersection of its diagonals.

**Solution.** In the given parallelogram $ABCD$, all four of its vertices are inaccessible. We have to locate the point of intersection of its diagonals.

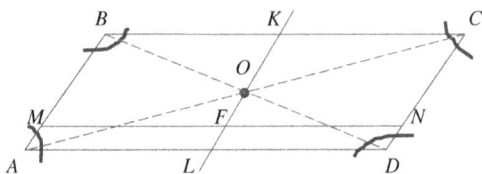

As evidenced a few times before, we will need to contemplate the plan for our solution based on the most important properties of the figure given in the problem.

For a parallelogram, clearly, it is the fact that the opposite sides are parallel and equal in size.

Even though we cannot use the inaccessible vertices of $ABCD$, we can use any point on its sides. Pick a random point $M$ on $AB$ and draw straight line $MN$ ($N$ lies on $DC$) parallel to $AD$ and $BC$. This is one of the basic constructions (draw a straight line through a point parallel to the given straight line; for this purpose, we can select any points on the straight line $AD$ other than $A$ and $D$) that we assume readers are familiar with and can easily complete using a compass and a straightedge. Having $MN$, we can find its midpoint $F$ (it is the second basic construction to be done by the readers) and draw straight line $KL$ ($K \in BC$, $L \in AD$) passing through $F$ parallel to $AB$ and $DC$. The final step will be to find the midpoint $O$ of the segment $KL$. This point has to be the desired point of intersection of $AC$ and $BD$.

Let's prove it. By construction, $F$ is the midpoint of $MN$, and $MN \parallel AD \parallel BC$. Therefore, we obtained two congruent parallelograms $AMFL$ and $LFND$ in which $MF = AL = FN = LD$. So, we see that $L$ is the midpoint of $AD$. In a similar fashion, we observe that $BK = MF = KC = FN$ and we can get that $K$ is the midpoint of $BC$. Consider now triangles $BOK$ and $DOL$. They are congruent by side-angle-side (SAS) because $BK = DL$, $KO = LO$, and $\angle BKO = \angle DLO$, as alternate interior angles of parallel sides $BK$ and $DL$ and transversal $LK$. It follows that $BO = DO$ as the corresponding sides in the congruent triangles. We proved that the midpoint of the segment $KL$ coincides with the midpoint of the diagonal $BD$. Since a segment can have only one midpoint, and we know that the diagonals of a parallelogram divide each other in half, this completes our proof.

It merits emphasizing that in our solution we did not even attempt to draw the diagonals passing through inaccessible vertices (this would require locating at least two accessible points on each) to find the point of their intersection. We merely utilized the definition and properties of a parallelogram to get the desired result. We found $O$ as the midpoint of the supplementary segment $KL$ and proved that it has to be the point of intersection of the diagonals as well.

Before we go on in our constructions' exploration, let's recall one important property of a trapezoid, which will be instrumental in many of our ensuing discussions. It was proved in Problem 9, Jacob Steiner's problem, in Chapter 6.

*In a trapezoid, the midpoints of the bases are collinear with the point of intersection of the diagonals, and also with the point of intersection of the two non-parallel sides.*

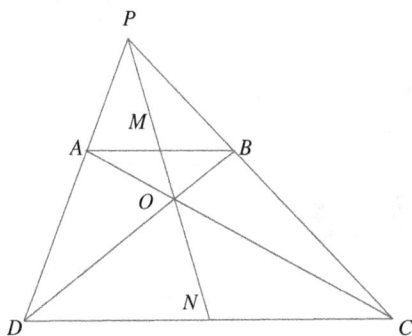

In the trapezoid $ABCD$, we draw its diagonals $AC$ and $BD$ intersecting at $O$ and extend the non-parallel sides $DA$ and $CB$ till their intersection at $P$. Straight line $PO$ passes through midpoints $M$ and $N$ of the respective parallel sides $AB$ and $DC$.

**Problem 2.** Given a triangle with two inaccessible vertices, draw its median from the third vertex.

**Solution.**

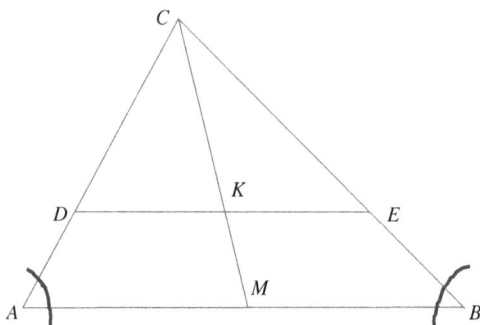

In the given triangle $ABC$, the vertices $A$ and $B$ are inaccessible. Our goal is to draw the median $CM$ to the side $AB$. Not being able to use vertices $A$ and $B$, we clearly can't find the midpoint $M$ of $AB$. The goal then will be to locate any point $K$ other than $C$, which belongs to our median drawn from $C$ to $AB$. As soon as we find such a point, we should be able to draw straight line $CK$. Its intersection with $AB$ will be the midpoint of $AB$, and the problem will be solved.

Even though we cannot use points $A$ and $B$, we can pick a random point $D$ on side $AC$ and draw $DE \parallel AB$, where $E$ belongs to $CB$. This is a standard construction of a straight line parallel to the given line passing through the given point. Now, we can find the midpoint $K$ of $DE$ and draw $CK$ till its intersection with $AB$ at $M$. By construction, $ADEB$ is a trapezoid, of which its extended non-parallel sides intersect at $C$. Since the midpoints of the bases in any trapezoid are collinear with the point of intersection of its non-parallel sides, $M$ has to be the midpoint of $AB$. Therefore, $CM$ is the desired median, and the problem is solved.

In our solution, the key decision was to draw a line parallel to the triangle's side with inaccessible vertices. It allowed us to introduce the auxiliary trapezoid, properties of which we exploited to solve the problem.

The suggested construction ideas utilized in Problem 2 prove to be helpful in solving the following problems 3 and 4 below.

**Problem 3.** Given a triangle with three inaccessible vertices, find its centroid.

**Solution.** The centroid of a triangle is the point of intersection of its medians. So, to find the point of intersection of the medians, we have to solve a more complicated problem than the previous one, because now all three vertices of the triangle are inaccessible. The intermediate step in our solution will be to figure out how to draw the median $CM$ in the given triangle from inaccessible vertex $C$ to the opposite side, the end points of which, $A$ and $B$, are inaccessible as well.

Not being able to use $C$, as we did in Problem 2, we have to find at least two points on our median to be able to draw it. This will be achieved by constructing two random trapezoids such that their vertices are located on the sides of our triangle and their parallel sides are parallel to $AB$. We draw $FL \parallel DE \parallel KN \parallel AB$ (see the below figure).

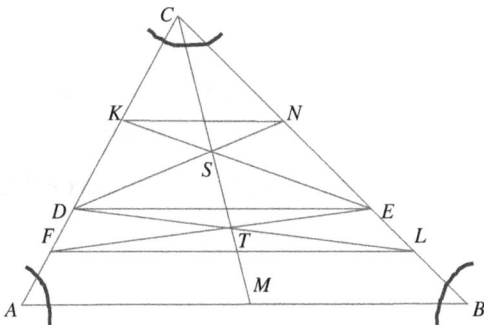

By doing this, we obtained trapezoids $DKNE$ and $FDEL$, parallel sides of which are parallel to $AB$ as well. Let $S$ be the point of intersection of the diagonals $DN$ and $EK$ of the first trapezoid and $T$ be the point of intersection of the diagonals $FE$ and $LD$ of the second trapezoid. These points lie on a straight line passing through the midpoint $M$ of $AB$ and point $C$, the triangle's inaccessible vertex, which is the point of intersection of the common non-parallel sides of the trapezoids. Straight line $ST$ is constructible, and it contains the median $CM$.

Let's turn now to our original problem and consider the following figure.

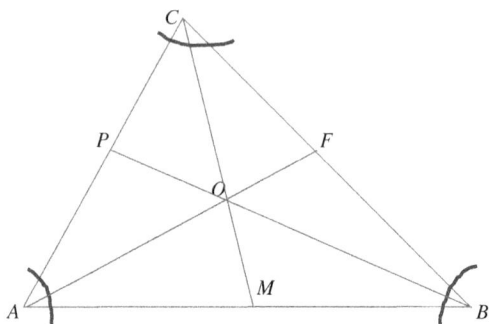

We are not going to show all the constructions in the above figure, and instead leave this for the readers. From the analysis above, it is clear that you have to repeat all the discussed steps and draw two pairs of arbitrary trapezoids with vertices on the sides of $\triangle ABC$ and with parallel sides parallel, respectively, to two sides of the given triangle. It will allow us to get the pair of points on each of the medians and consequently, we can draw the medians through them. There is no need to draw the third median; it must pass through their point of intersection as well. Point $O$, as the point of intersection of the triangle's medians, is the desired centroid.

**Problem 4.** Given a triangle with three inaccessible vertices, find its circumcenter.

**Solution.**

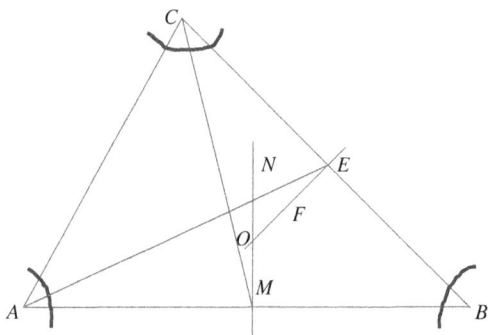

The circumcenter of a triangle is the point of intersection of perpendicular bisectors to its sides. Hence, we need to first find the midpoint of each side, and then draw the perpendicular to each side through its midpoint till their intersection. The first part can be solved by following the steps outlined in Problem 3 for construction of the triangle's medians $CM$ and $AE$. The next step will be to draw $MN \perp AB$ and $FE \perp BC$. Points $A$, $B$, and $C$ are inaccessible, but we can locate any other pair of points equidistant from the found midpoint on each side, and then do twice the basic construction of the line perpendicular to the given line passing through the given point on that line. Once again, we leave these simple exercises to readers. The point of intersection of $MN$ and $EF$, point $O$, is the circumcenter of the given triangle.

As we see, the solutions of Problems 2, 3, and 4 rely on the same interesting property of a trapezoid. The solutions appear to be closely related to each other after we consider them one after the other. Imagine now that you face one of these problems, especially Problem 3 or Problem 4, not as in the context above, but independently; let's say at some math contest. What is the essential idea to approach each problem? The hint is in exploiting an auxiliary trapezoid and its properties. It will allow for identifying and constructing the medians, which finally will lead to the desired solution.

**Problem 5.** Draw a perpendicular through the inaccessible point of intersection of the two given straight lines to the third given straight line, which intersects them.

**Solution.** In fact, we are dealing here again with a triangle with an inaccessible vertex. By rephrasing the problem, we can simplify our search for the solution. This problem is about constructing the altitude of a triangle $ABC$ from inaccessible vertex $C$ to the opposite side $AB$.

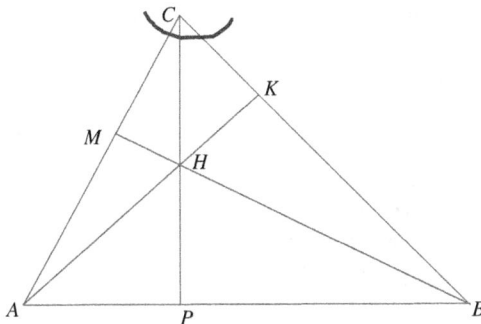

So, the question is if we can find a point on the altitude drawn from $C$ other than its vertex. If yes, then the problem converts to a standard construction of a perpendicular from the given point to the given line. We know that the three altitudes in a triangle intersect at one point, its orthocenter. This point in our case can be easily found as the intersection of altitudes $AK$ and $BM$ dropped to the respective sides $BC$ and $AC$. Don't be confused by the fact that $C$ is not usable. We can locate any other points on $BC$ and $AC$ and then do the constructions of our altitudes. This can be done in a two-step process; by selecting any two points on each straight line, drawing a perpendicular to the selected segment, and then drawing the straight line through $A$ and $B$ parallel to those perpendiculars till the intersection with each opposite side. Getting $H$, the orthocenter of triangle $ABC$, we finally have to drop the perpendicular from $H$ to $AB$, $HP \perp AB$. The third altitude of $ABC$ is contained on this perpendicular; therefore, $HP$ will pass through $C$ and is the desired solution.

**Problem 6.** Given a triangle with three inaccessible vertices, find its orthocenter.

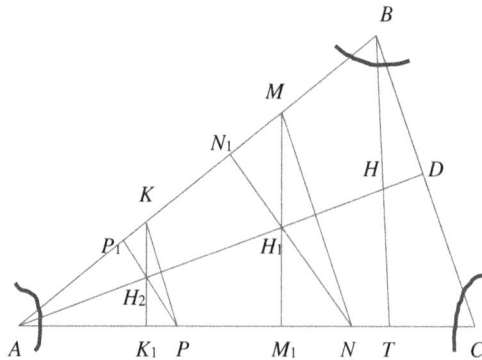

**Solution.** Our goal in this problem is to draw two altitudes and find the point of their intersection (the third altitude will pass through this point as well). So, now, we deal with the construction of a perpendicular from an inaccessible point to a segment with inaccessible end points. The obvious choice will be to utilize again the fact of the orthocenter's existence as the point of intersection of the three altitudes in a triangle and apply the results obtained in Problem 5. Pick any point $M(M \neq A, M \neq B)$ on $AB$ and draw $MN \parallel BC$. Then, draw $MM_1 \perp AC$ and $NN_1 \perp AB$. Point $H_1$ of their intersection is the orthocenter of triangle $AMN$. In a similar fashion, we pick a random point $K(K \neq M)$ on $AB$ and draw $KP \parallel MN \parallel BC$ and $KK_1 \perp AC$ and $PP_1 \perp AB$. Point $H_2$ of the intersection of $KK_1$ and $PP_1$ is the orthocenter of triangle $AKP$. Since by construction, $KP \parallel MN \parallel BC$, both points $H_1$ and $H_2$ must be located on the altitude of the given triangle $ABC$ dropped from $A$ to $BC$. Therefore, $H_1 H_2$ must pass through $A$ and is perpendicular to $BC$ (another option was just to draw the perpendicular to $BC$ through $H_1$). Denote $D$ the point of its intersection with $BC$. Then, $AD$ is the first altitude drawn in triangle $ABC$. In a similar manner, we can draw the second altitude $BT$. We omit the details here, but suggest the readers do all the steps and finish the construction. Finding $H$, the point of intersection of altitudes $AD$ and $BT$, we conclude our solution.

In the above problems, we were able to locate the classic centers — centroid, circumcenter, and orthocenter in a triangle with inaccessible vertices. Another similar interesting construction problem is to find the center of the inscribed circle in a triangle with three inaccessible vertices. We are not going to present it here because the detailed solution was given in my book *Geometrical Kaleidoscope*, Dover Publications, 2017. So, one can find it there or it should be a good exercise for the ambitious reader to find the solution on his/her own.

I find it amazing that even though all the considered construction problems look difficult and non-trivial at first, essentially, the major key to the solution of each was merely to utilize the fact of each classic center's existence.

It is worth pointing out here another compelling solution for Problems 3, 4, and 6 using the results of any two out of the three problems to solve the remaining one. It can be done by applying the remarkable property that the circumcenter, orthocenter, and centroid of a triangle lie on the same straight line, the *Euler line*, with the centroid dividing the distance from the orthocenter to the circumcenter in the ratio 2:1.

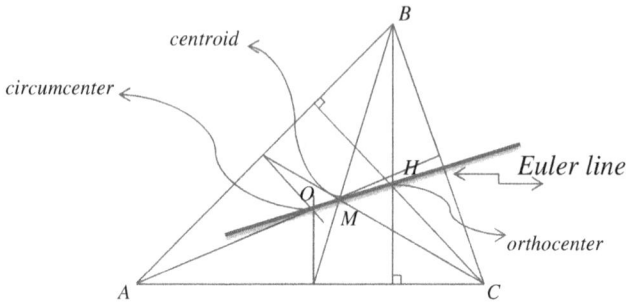

Assume, for example, that we indeed have already solved the construction problems for locating the centroid and circumcenter of a triangle with inaccessible vertices. To find its orthocenter, we merely have to locate the point on the Euler line that connects both found points, such that its distance from the centroid is twice the distance between the centroid and the circumcenter. In a similar fashion, we can solve the other two problems of locating the centroid and circumcenter assuming that the respective problems of constructing the orthocenter and circumcenter, and orthocenter and centroid are already solved.

Let's now consider a more complicated problem, the solution of which heavily relies on the problems considered above.

**Problem 7.** Given a trapezoid $ABCD$ with four inaccessible vertices, locate the point of intersection of its diagonals.

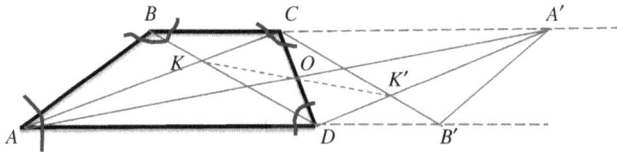

**Solution.** To solve this problem, we suggest considering the properties of the central symmetry and do two half-turns to find the image of $ABCD$ that will have all accessible vertices. Having the congruent trapezoid image with four accessible vertices, we will be able to find the point of intersection of the diagonals of the image. By doing then two consecutive constructions of a reverse image of that point in our central symmetries, we will be able to locate the desired point of intersection of the diagonals of the original trapezoid.

We will provide the detailed analysis of the solution leaving the actual constructions to be done by the readers.

First, we will consider a half-turn with the center at the midpoint of $CD$ point $O$. To locate $O$, we will work with triangle $ACD$ with three inaccessible vertices and draw its median $AO$. This construction was explained in Problem 3. Next, consider the central symmetry with the center $O$ and find the image of inaccessible point $A$ (see the above figure). To find such an image, the point $A'$, we have to find the point of intersection of the extensions of $BC$ and $AO$. This can be justified by the following observations.

Since $A'$ lies on extension of $BC$, and $BC \parallel AD$, then $CA' \parallel AD$. It follows that $\angle ADO = \angle A'CO$ as alternate interior angles formed by two parallel lines and transversal $CD$. In triangles $AOD$ and $A'OC$, we also have $CO = DO$ by construction, and $\angle AOD = \angle A'OC$ as vertical. Therefore, these two triangles are congruent by (angle-side-angle (ASA). Thus, the other pair of the respective sides is of the same length, $CA' = AD$ and $A'O = AO$. So, indeed, $A'$ is the image of $A$ in the considered point symmetry. $O$ divides both segments $CD$ and $AA'$ in half. We see that as the result of our constructions we obtained the parallelogram $ACA'D$. Similarly, we can find $B'$, the image of $B$ in the same point symmetry with the center $O$. $BCB'D$ is the parallelogram as well. In our point symmetry with the center $O$, segment $AB$ was transferred into $A'B'$, $BC$ into $B'D$, $AD$ into $A'C$, and $CD$ into $DC$. Hence, the trapezoid $ABCD$ was transferred into the congruent trapezoid $A'B'DC$. We obtained the new trapezoid which has two inaccessible vertices $C$ and $D$, and two accessible vertices $A'$ and $B'$.

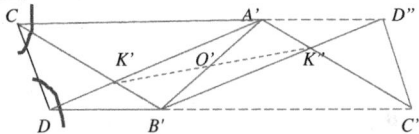

If we repeat the described constructions one more time, now constructing the image of $A'B'DC$ in central symmetry with the center at midpoint $O'$ of $A'B'$, we should get another trapezoid $A'D''C''B'$ congruent to $B'DCA'$ (see the above figure). All the vertices of the last trapezoid are accessible. By drawing the diagonals $A'C''$ and $B'D''$, we locate their point of intersection $K''$. As we locate $K''$, we can construct its reverse image in the central symmetry with the center $O'$ and find $K'$, the point of intersection of $A'D$ and $B'C$. Finally, going back to the figure given under Problem 7, we can construct $K$, the point of intersection of the diagonals $AC$ and $BD$, as the reverse image of $K'$ in the central symmetry with the center $O$.

The decision to apply central symmetry was instrumental in solving Problem 7. Analyzing this solution, we should admire the fact that Euclidean plane transformations present a rewarding alternative for the solution of all of the problems for locating the orthocenter, centroid, and the incenter, point of intersection of angle bisectors in a triangle with inaccessible vertices. Indeed, you can apply any of the isometric transformations to a figure with inaccessible elements, construct its image in a specific transformation, find the required element on the new figure, and make a reversal transformation to the original to find the reverse image of the found element, a similar approach as we used in our solution of Problem 7.

Let's demonstrate this with the following alternative solution, for instance, of Problem 4 for the construction of a circumcenter of a triangle with three inaccessible vertices.

**Solution.** Let's select a random point $S$ outside of triangle $ABC$ and construct the image of $ABC$ in a half-turn with respect to center $S$ (the same as rotation by $180°$ with respect to $S$).

Since $A$, $B$, and $C$ are inaccessible, we locate any other two points $E$ and $K$ on $AB$ and find their images $E_1$ and $K_1$ in this central symmetry. Draw a straight line through $E_1$ and $K_1$. It has to be the line containing the image of $AB$. Select any two points on $BC$ and construct their images in the same half-turn with respect to $S$; for simplicity you can select as these two points the points of intersection of $ES$ and $KS$ with $BC$.

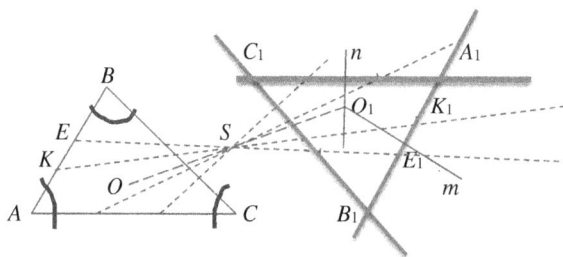

Draw a straight line through the images of these two points; it is the line containing the image of $BC$. Repeat the construction for any two random points on $AC$ and find their images; draw a straight line through them. This straight line contains the image of $AC$. Finding the points of intersection of the pairs of the constructed images of $AB$, $BC$, and $AC$, we get the vertices $A_1$, $B_1$, and $C_1$ of the triangle $A_1B_1C_1$, the image of $ABC$ in the considered half-turn with respect to $S$. All the elements of the new triangle $A_1B_1C_1$ are accessible and we can now easily locate the circumcenter of $A_1B_1C_1$, point $O_1$, as the point of intersection of the perpendicular bisectors to its sides, $n$ and $m$. As we get $O_1$, the final step will be to construct the reverse image of $O_1$ in the half-turn with respect to $S$, point $O$, which will be the desired circumcenter of $ABC$.

Considering rotation by any other than a 180° angle, translation, reflection, or even a homothety would provide the same result. We will be able to construct the image of $ABC$, triangle $A_1B_1C_1$, find its circumcenter, and locate its reverse image in a specific transformation. This point will be the desired circumcenter of the original triangle with inaccessible vertices. It should be a good exercise for ambitious readers to do all the mentioned constructions not just for a circumcenter but for the other classic centers of a triangle as well.

**Problem 8.** Given a circle and the inaccessible point outside of that circle as the intersection of the two given straight lines, construct two tangent lines through this point to the circle.

**Solution.** Let's first analyze what is given in this problem. We have the circle with the center $O$ and two straight lines $AM$ and $AP$ intersecting at inaccessible point $A$. In other words, we can't use $A$ in any of our constructions; however, any other point on $AM$ and $AP$ is available for us. The goal is to draw two tangents from inaccessible point $A$ to the given circle.

Knowing that a tangent from a point outside of a circle is perpendicular to a radius dropped to the point of tangency, we need to locate somehow such points of tangency on the given circle.

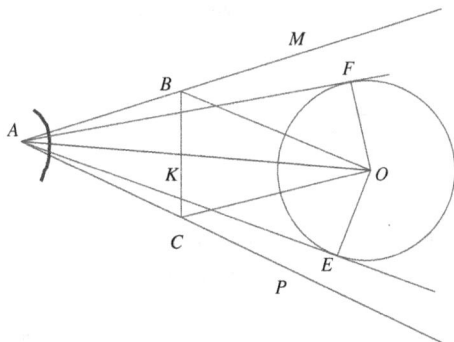

If we manage to do this, our next step will be to draw a perpendicular through each such point to the radius connecting them to the circle's center $O$, which is doable standard construction. Assume that our problem is solved, and the points of tangency are $F$ and $E$. Triangles $AFO$ and $AEO$ are congruent right triangles because $AO$ is the common hypotenuse and $OF = OE$ as radii of the same circle. Therefore, both triangles have to be inscribed in the same circle with the center at the midpoint of the common hypotenuse and radius equal to half of the hypotenuse (we will be discussing this property in detail in Chapter 9). So, our problem is now reduced to finding the midpoint of $AO$. Let's draw $OB \parallel AP$, $B \in AM$ and $OC \parallel AM$, $C \in AP$. By construction, $ABOC$ is a parallelogram. We know that the diagonals in a parallelogram bisect each other. Therefore, by finding $K$, the midpoint of $BC$, we get the point of intersection of the diagonals in $ABOC$. This midpoint $K$ is easily constructible for the obtained segment $BC$. The next step is to draw a circle with the center $K$ and radius equal to $KO$. The points of intersection of this circle with the given circle will be the desired points of tangency $F$ and $E$. Indeed, since $K$ is the midpoint of $AO$, the second diagonal of our supplementary parallelogram $ABOC$, and by construction, $KA = KO = KF = KE$, all four points, $A$, $F$, $O$, and $E$ lie on the circle with the center $K$ and radius equal to the distance from $K$ to each of these four points. Each of the inscribed angles $AFO$ and $AEO$ is subtended by the diameter $AO$. Hence, $\angle AFO = \angle AEO = 90°$. It implies that $F$ and $E$ are the desired points of tangency. The final step is to draw lines perpendicular to $OF$ and $OE$ through $F$ and $E$, respectively. These perpendiculars will be the desired tangents to the given circle passing through $A$.

Construction puzzles with restricted elements are especially attractive because one might face those problems in real-life situations. For example, such situations may occur in constructing a building, in geological field work, archeological investigations, or in the land-surveying and geodesy industry.

To conclude this chapter, we will offer one of those real-life problems related to measuring the distances between the given point and an inaccessible point.

**Problem 9.** Measure the distance between the house and the tower located on the different banks of a river assuming you can draw straight lines between accessible and inaccessible points on different banks of the river by using, for example, geodesic equipment, but cannot physically measure the distances between them.

**Solution.** First, let's clearly understand what we are dealing with and what is requested of us to be done. We have two points $A$ (the house) and $B$ (the tower) separated by the river. We can draw straight lines connecting $A$ and any point on our bank of the river, where point $B$ is located. The goal is to measure the distance between $A$ and $B$, or in other words, design some plan that would allow us to construct on our bank of the river a segment with a length equal to $AB$, so we can measure it.

This problem is different from all the considered problems in which inaccessible points were not available for our use at all. We can draw now straight lines (position them on the sketch) connecting $A$ with points on the other bank of the river. However, we cannot do measurements between $A$ and any point located on the other bank. Not being able to access point $A$ on the other bank of the river, we though have available a portion of the segment connecting $A$ and $B$ on our bank of the river, each point of which can be utilized in our constructions.

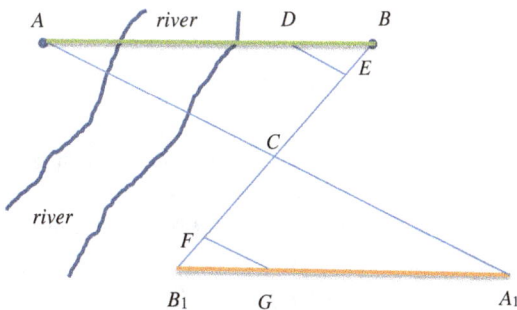

If we manage to construct two congruent triangles such that they have one of the respective sides with the length of $AB$, with one of them completely located on our bank of the river, our goal will be achieved.

Select a random point $B_1$ on our bank of the river and draw a straight line through $B_1$ and $B$. Locate the midpoint $C$ of $B_1B$. Select random points $D$ located on the accessible side of the segment $AB$ and $E$ on $B_1B$ and connect them. We can measure now the distance between $D$ and $E$ and $B$ and $E$, and construct the segment $FG$ such that $FG \parallel DE$, $FG = DE$, with $F$ located on $B_1B$ and $B_1F = BE$. Draw a straight line through $B_1$ and $G$ and locate point $A_1$ on it, such that it belongs to $AC$ as well. Segment $A_1B_1$ has the same length as $AB$ and it is measurable because it is within our reach being located on the same bank of the river as we are.

Let's prove now that indeed $A_1B_1 = AB$.

By construction, $B_1F = BE$, $FG = DE$, and $\angle B_1FG = \angle BED$ as alternate interior angles by parallel lines $FG$ and $DE$ and transversal $B_1B$. Therefore, triangles $B_1FG$ and $BED$ are congruent. It follows that $\angle FB_1G = \angle EBD$ as the corresponding angles in the congruent triangles. This implies that $\angle CB_1A_1 = \angle CBA$.

So, we get that in triangles $ACB$ and $A_1CB_1$ the corresponding sides and two adjacent angles are equal, $B_1C = BC$, $\angle ACB = \angle A_1CB_1$ as vertical, and $\angle CB_1A_1 = \angle CBA$, as proved above. Hence, these triangles are congruent, and it follows that their respective sides $AB$ and $A_1B_1$ have the same length. Our proof is completed, and we conclude that indeed, we managed to get the measurable segment $A_1B_1 = AB$ on our bank of the river.

This interesting problem can be slightly modified if instead of permission to use geodesic equipment to draw straight lines between points located on the different banks of the river, we have some other object, for example, a point $B_1$ on the same bank of the river as $B$, and there is available a portion of the segment connecting $A$ and $B$ and $A$ and $B_1$ on our bank of the river.

So, in this modified problem, we cannot draw straight lines using inaccessible point $A$, but we can use the portions of sides $BA$ and $B_1A$ of the angle $BAB_1$. Considering the same plan for solving this modified version (all the same steps in drawing $DBE$ and $GB_1F$), we will be relying on the result of Problem 3 solved at the beginning of this chapter explaining how to draw the median from an inaccessible vertex of a triangle to the opposite side.

Indeed, having the specific location of $B$ and $B_1$ and portions of the sides of the angle $BAB_1$ located on our bank of the river, we will need to draw the median $AC$ in the triangle $BAB_1$, the extension of which gives $A_1$ as the intersection with $B_1G$.

It should be good practice for ambitious readers to complete the solution of this modified version, justifying all the steps and analyzing the results; maybe you could even invent your own new problems involving constructions and measurements with inaccessible points.

## Chapter 8

# Inventing a Problem

"Practice makes perfect" — this is the best recipe for becoming a successful math problem solver. The more problems one solves, the more proficient he/she becomes, the easier he/she recognizes families of related problems, solutions of which are based on the same or similar principals. But, mathematics isn't about learning a collection of facts — it is about learning to think logically and creatively. Often, the ideas utilized in the solving of certain problems become useful in other, not directly related problems. Raising questions and making conclusions, analogies, or generalizations from the obtained results lead to designing your own problems. This greatly stimulates creativity and deeper understanding of properties and relationships studied. Moreover, this may lead to new amazing discoveries. In this and the next chapter, we will demonstrate how to generate new problems from some basic and classic properties, "invent" new relationships, and get interesting results from the questions raised. This can be accomplished, for example, through the following different approaches:

1. Use a solved problem or a specific property or a theorem and build your own problem based on the results obtained from it. Manipulate variations of your outcome to make the new problem interesting.
2. Consider a "basic" problem and see if you can devise a similar example with a set of parameters to produce a predictable result.
3. Raise questions from some unexpected or bypass results achieved in a process of a specific problem solving. Your inquiries may

lead to inventing new interesting (not even necessarily solvable) challenges not directly related to the problem you were solving at the beginning.

Let's now go through examples illustrating each of the above brainstorming approaches in new problem improvisations.

## 8.1   Inventing New Problems by Modifying a Solved Problem

**Problem 1.** Construct a segment with the given length that is passing through the point of intersection of the two given circles, such that its end points are on each circle.

**Solution.** This construction calls for a segment $MN$, the length of which is given, to pass through $S$, the point of intersection of the two given circles with the centers $O_1$ and $O_2$, such that $M$ is located on the first circle and $N$ is located on the second one.

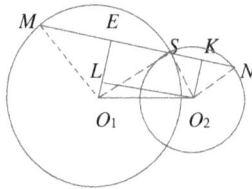

Assume the problem is solved. Let's do a few additional constructions and analyze what we have.

First, draw $O_1E \perp MN$ and $O_2K \perp MN$. Since $O_1M = O_1S$ as the radii of the first circle, then triangle $MO_1S$ is isosceles. $O_1E$ being an altitude dropped to the base of an isosceles triangle is its median at the same time. Therefore, $ME = ES$. In a similar fashion, considering the isosceles triangle $NO_2S$, we conclude that $SK = KN$. It follows that

$$EK = ES + SK = \frac{1}{2}MS + \frac{1}{2}SN = \frac{1}{2}(MS + SN) = \frac{1}{2}MN.$$

Now, draw $O_2L \parallel EK$. We obtain the rectangle $LEKO_2$ in which $LO_2 = EK = \frac{1}{2}MN$. So, in the right triangle $O_1LO_2$ ($\angle L = 90°$), we know the length of the leg $O_2L$ and the length of the hypotenuse $O_1O_2$ (given both circles with the centers $O_1$ and $O_2$, clearly, we know the length of $O_1O_2$). Recall that in order to construct a triangle with a straightedge and compass, we require the measure of three parts of a triangle. Hence, the right triangle $O_1LO_2$ is constructible. After constructing $\triangle O_1LO_2$, we can draw a straight line through point $S$ parallel to $LO_2$. It will intersect the circles at points $M$ and $N$. Segment $MN$ will have the given length, and therefore it will satisfy the problem's conditions.

So, our constructions steps are as follows:

(1) Connect $O_1$ and $O_2$.
(2) Construct the right triangle $O_1LO_2$ such that $O_2L = \frac{1}{2}MN$ and $\angle L = 90°$.
(3) Draw a straight line containing $S$, parallel to $LO_2$, and intersecting each circle at $M$ and $N$, respectively. $MN$ is the desired segment.

We leave the actual constructions and justifications of the steps to the readers.

We can easily envision new problems by varying the data of the original problem. Looking back at our analysis and the solution's steps, and adding some new notion to the solved problem, we should be able to pose a few natural questions:

(1) Nothing specific was said about the length of $MN$. So, out of all such segments satisfying the problem's conditions, which one will have the maximum length? How can it be constructed?
(2) Is it possible to draw $MN$ through $S$ in such a way that $MS = SN$?
(3) If we manage to find an answer to the previous question, can we extend the problem to a more general case and draw $MN$ through $S$, in such a way that $S$ divides $MN$ in a specific ratio, $\frac{MS}{SN} = \frac{m}{n}$ ($m$ and $n$ are some natural numbers)?

In fact, by making the above inquiries, we just formulated three more interesting challenges and invented new problems, solutions of which are worth considering here.

**Problem 1.1.** Given two intersecting circles with the centers $O_1$ and $O_2$, draw a segment through the point of their intersection $S$, such that its end points are on each circle and it has the maximum length out of all such segments.

**Solution.** Referring to the solution of our original problem, it should be clear that the segment with the maximum length satisfying the problem's conditions has to be parallel to $O_1O_2$. Indeed, the hypotenuse is the longest side of a right triangle. So, with our leg $O_2L$ "approaching" the hypotenuse $O_1O_2$, the length of $MN = 2O_2L$ will be maximized when it becomes $MN = 2O_1O_2$. This will be achieved when we draw $MN$ through $S$ parallel to $O_1O_2$.

**Problem 1.2.** Given two intersecting circles with the centers $O_1$ and $O_2$, draw a segment $MN$ through the point of their intersection $S$, such that its end points are on each circle and $MS = SN$.

**Solution.** This variation of the original problem presents a more difficult challenge than the previous one.

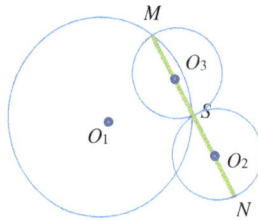

A compelling and elegant solution will be achieved by applying the properties of central symmetry. Construct an image of the circle with the center $O_2$ in the central symmetry with the center $S$. Such an image will be the identical circle with the center $O_3$. Denote by $M$ the point of its intersection with the circle with the center $O_1$. By construction, points $M$, $O_3$, $S$, and $O_2$ are located on the same straight line, and $MO_3 = O_3S = SO_2$ as radii of the identical circles (since central symmetry preserves the distances, the radii of both circles with the centers $O_3$ and $O_2$ are equal). Extend this straight line till its point of intersection $N$ with the circle with the center $O_2$. This point $N$ will be the desired second end of our segment $MN$, such that $MS = SN$. Indeed, these segments are equal as the diameters of our identical circles with the centers $O_3$ and $O_2$, and $M$ and $N$ lie on the respective given circles with the centers $O_1$ and $O_2$. Hence, $MN$ satisfies the conditions of the problem and is the desired segment.

**Problem 1.3.** Given two intersecting circles with the centers $O_1$ and $O_2$, draw a segment $MN$ through the point of their intersection

$S$, such that its end points are on each circle and $\frac{MS}{SN} = \frac{m}{n}$ ($m$ and $n$ are some natural numbers).

**Solution.** Utilizing our solution from the previous problem and recalling the properties of a homothety, we should be able to get the desired result in a similar fashion as we did by applying central symmetry in Problem 1.2. In this case, we will need to construct an image of the circle with the center $O_2$ in a homothety with the center $S$ and ratio $k = \frac{m}{n}$. We will obtain point $M$ as the point of intersection of the circle with the center $O_3$ (the homothetic image of the circle with the center $O_2$ in our homothety) with the circle with the center $O_1$. Then, we will get $N$ as the point of intersection of the straight line passing through the center of homothety $S$ and point $M$ and the circle with the center $O_2$. $MN$ is the desired segment because $\frac{MS}{SN} = \frac{m}{n}$ (by construction, according with homothetic properties) and its end points are located on the given two circles. We leave all the constructions and justifications of steps to the readers to complete.

As you can see, while solving three interesting modifications of the original problem, we got positive responses to all the raised questions. We should note that by recalling the properties of rotations and homothety, we were able to see how appealing these properties are in finding elegant and relatively simple solutions for construction problems.

I think it was not hard to "invent" three of the above variations of the original problem; indeed, we just slightly modified one of the given attributes. It takes much more to invent a new interesting problem relying on the obtained results, which you can't relate *directly* to the solved problem. It requires ingenuity and inventiveness to hide the "basic" problem inside your new problem and make it appealing and challenging; it is not an easy task at all.

At some degree, I would compare the inventing of a new problem to masterminding a detective story. A masterfully designed detective story keeps you on the edge of your seat. The author provides you with the facts and details, and describes the circumstances and main characters incorporated in the plot of a committed crime (given original conditions in a problem). The story intrigues you into asking questions on how the crime was committed and who is to blame for it (what has to be proved or found in a problem, its question).

The criminal is usually not revealed up until the last pages of the novel, and his revelation often turns out to be the most unexpected part of the story. Even though it looks like you know enough information to be able to deduce the criminal yourself, your solution does not always coincide with author's final outcome. The better the author camouflages the reasons and ways to disclose the story's ending, the more beguiling and captivating it appears. It is very similar to inventing a new problem having in mind some specific property or already solved problem, which is intended to be the backbone for a solution. If you manage to hide your main idea, and conceal or even masquerade the obvious path to the solution, the more challenging and interesting the problem can become.

We suggest the readers try the following problem to solve before reading the suggested solution. You do have a hint that Problem 1.4 is somehow related to our original Problem 1; the hint is in its number. I think it would be interesting to compare your results to our solution below.

**Problem 1.4.** Inscribe the given triangle $MNP$ into the given triangle $ABC$.

**Solution.** To inscribe $\triangle MNP$ into $\triangle ABC$ means to construct triangle $MNP$ with three sides that have the given lengths and three angles that have the given values in such a way, that every vertex of $\triangle MNP$ is located on one side of the given $\triangle ABC$.

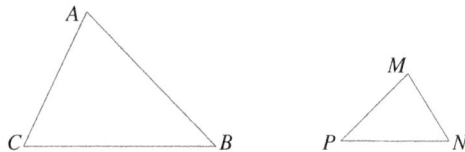

This problem presents a good example of the case when it is beneficial instead of solving the original problem to consider the restated equivalent problem:

*Circumscribe the given triangle $ABC$ over the given triangle $MNP$.*

At first glance, the problem we are dealing with has nothing to do with Problem 1. Even after restating the problem, it's not clear how it can be related to the already solved problem we are referring to.

However, by restating the problem we point out the direction to finding the path for its solution (in the next chapter, we will be talking more about this helpful technique of restating a problem and changing the focus of contemplating the plan for its solution). Instead of constructing $\triangle MNP$ inscribed in the given $\triangle ABC$, we will try to construct the triangle congruent to $\triangle ABC$ circumscribed over $\triangle MNP$.

So, let's analyze the problem and see if we can come up with a plan for the solution. First of all, notice that we are given all the lengths of the sides and the angle measures of both triangles $\triangle MNP$ and $\triangle ABC$. It implies that we can easily determine the radius of each circle circumscribed over triangles $MAN$ and $NPB$.

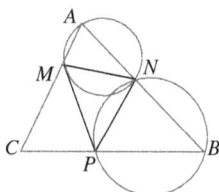

Knowing the length of $MN$ and the measure of angle $A$, by applying the *Law of sines*, we can find the radius of the circumcircle of the triangle $MAN$ as $r_1 = \frac{MN}{2\sin\angle MAN}$. It follows that we can draw a circle with the radius of the length equal to $r_1$ circumscribed over $\triangle MAN$. Similarly, we can find the radius of the circumcircle of the triangle $NPB$ as $r_2 = \frac{PN}{2\sin\angle PBN}$, and draw a circle circumscribed over $\triangle NPB$.

At this moment, you should appreciate the result obtained during the solution of the original Problem 1. Indeed, when we started working with $\triangle MNP$, we realized that we can construct two circumscribed circles over triangles $MAN$ and $NPB$. They will intersect at point $N$. So, our problem is reduced to drawing the segment with the length equal to $AB$ through point $N$, such that $A$ and $B$ are located on each of the circumcircles of the triangles $MAN$ and $NPB$, respectively, i.e., we face now exactly the same problem as we started with our "inventions". As soon as we locate points $A$ and $B$, we will draw $BP$ and $AM$ till their intersection at $C$. $\triangle ABC$ is the desired triangle circumscribed over $\triangle MNP$. We finalized the analysis of the problem, so the constructions steps are as follows (we suggest the readers thoroughly perform every basic construction, explanations of which we will omit here):

1. Start with the given $\triangle MNP$ and determine the length of each radius $r_1 = \frac{MN}{2\sin\angle MAN}$ and $r_2 = \frac{PN}{2\sin\angle PBN}$.
2. Draw a perpendicular bisector to segment $MN$ and locate on it the point equidistant from $M$ and $N$ at $r_1$. This will be the center of the first circle we need to draw.
3. Draw the first circle with the center located in Step 2 and the radius $r_1$, such that $M$ and $N$ lie on it.
4. Draw a perpendicular bisector to segment $PN$ and locate on it the point equidistant from $P$ and $N$ at $r_2$. This will be the center of the second circle we need to draw.
5. Draw the second circle with the center located in Step 4 and the radius $r_2$, such that $P$ and $N$ lie on it.
6. Draw a segment through $N$, such that it has the length equal to the side $AB$ of the given triangle $ABC$, and points $A$ and $B$ are located on each of our circles with the radii $r_1$ and $r_2$ (see Problem 1).
7. Draw $AM$ and $BP$ till their intersection at $C$. This concludes our construction of the given triangle $ABC$ circumscribed over the given triangle $MNP$.

At the beginning of our investigation of Problem 1.4, it was not apparent how it can be related to Problem 1. The problem looked difficult. However, analyzing the given conditions and changing the focus from inscribing $\triangle MNP$ into $\triangle ABC$ to do the task of circumscribing $\triangle ABC$ over $\triangle MNP$ (exactly the same problem, but allowing to look at it from a different perspective), we came across constructing the two intersecting circles at vertex $N$ of the given triangle $MNP$, and reduced our problem to Problem 1.

This mathematical "detective story" indeed had a happy ending in getting an elegant construction satisfying the given conditions. We suggest that readers investigate how many solutions this problem might have depending on the given attributes.

Let's now turn to and investigate one more intriguing mathematical "detective story". We have been referring several times in the course of the book to the problem about the properties of a line passing through the point of intersection of non-parallel sides of a trapezoid and the point of intersection of its diagonals to divide in half the trapezoid's bases. Let's demonstrate in the next problem how this property may be used in inventing new and interesting challenges that are not directly, at least from the first view, related to it.

**Problem 2.** Two triangles formed by the extended sides of a convex quadrilateral until their intersection have equal areas. Prove that one of the diagonals of this quadrilateral divides the other diagonal in half.

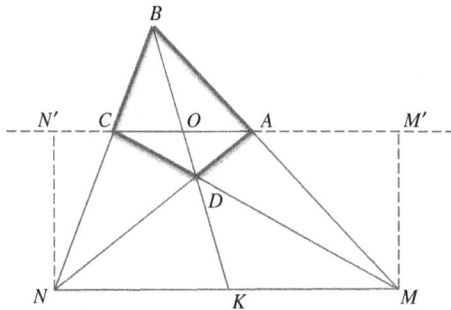

**Solution.** In the given convex quadrilateral $ABCD$, we extend $BA$ and $CD$ until the point of their intersection $M$, and we extend $BC$ and $AD$ until their intersection at $N$. We obtained two triangles $BNA$ and $BMC$ with equal areas. The goal is to prove that $BD$ divides $AC$ in half, i.e., the point of their intersection $O$ is the midpoint of $AC$.

This is the great example of a problem, solving which the best advice should be to "do whatever is doable analyzing the given conditions". Indeed, it is not clear how to get to the requested statement that $BD$ divides $AC$ in half. So, we will start from analyzing what we have. We know that areas of $BNA$ and $BMC$ are equal. This implies that the areas of triangles $DNC$ and $DMA$ are equal as well, because triangles $BNA$ and $BMC$ consist of these two triangles and of the common part $ABCD$. From this, we can conclude that the areas of triangles $ANC$ and $CMA$ have to be equal. Indeed, triangle $ADC$ is the common part of both triangles $ANC$ and $CMA$, and as proved, areas of $DNC$ and $DMA$ are equal. So, their sums with the area of $ADC$ are equal as well.

The area of a triangle equals half the product of its base by the altitude dropped to that base. Let's draw the altitudes $NN'$ and $MM'$

in triangles $ANC$ and $CMA$ ($\angle C$ and $\angle A$ are the obtuse angles in each triangle, so the foot of each altitude will be located on the extension of $AC$ to the left and to the right, respectively), and compare the areas:

$$S_{\triangle ANC} = \frac{1}{2} AC \cdot NN',$$

$$S_{\triangle CMA} = \frac{1}{2} AC \cdot MM'.$$

Since $S_{\triangle ANC} = S_{\triangle CMA}$, then obviously, $NN' = MM'$. So, the perpendiculars dropped from $M$ and $N$ to $AC$ are of the same length. Hence, $NN'M'M$ is a rectangle. It implies that its opposite sides located on the straight lines $MN$ and $AC$ are parallel. We see that as the result of our constructions, the opposite sides in the quadrilateral $NCAM$ are parallel and, therefore, it is a trapezoid. Now, having identified the auxiliary trapezoid, we can relate to the property that a line passing through the point of intersection of non-parallel sides of a trapezoid and the point of intersection of its diagonals divides in half the trapezoid's bases. $BD$ is such a line in our constructions, so it has to divide $AC$ in half!

While inventing this problem, we had in mind the mentioned property. It was well camouflaged, so after assessing the problem, at the beginning it was not clear how to use it in conceiving the plan for the proof. In fact, it came into sight and became available for discussion only when the auxiliary trapezoid had been introduced. It immediately enlightened the whole picture and allowed us to make the final conclusion in our proof.

## 8.2    Inventing New Problems by Analogy

We may derive many interesting problems by analogy.

One of my favorite classic geometric problems is Napoleon's Theorem:

> *The centers of equilateral triangles constructed on the sides of a triangle are the vertices of a regular triangle.*

$ABC$ is a triangle, on the sides of which the three regular triangles are constructed. Their centers form a regular triangle $O_1O_2O_3$.

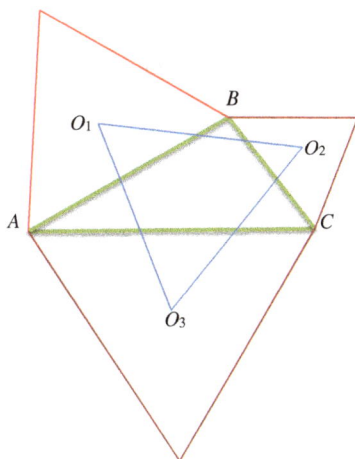

We are not going to solve this problem in the book. It has several interesting solutions, one of which by applying rotation properties was demonstrated in my book *Geometrical Kaleidoscope*, Dover Publications, 2017. We suggest the readers try the problem or read its solution, for example, in the mentioned book.

Instead, we are going to use Napoleon's Theorem to ponder some interesting questions, and create problems of our own that resemble it.

**Problem 3.** What type of a quadrilateral is formed by connecting the centers of the squares constructed on the sides of a rectangle?

**Solution.** This is relatively simple problem and it is not hard to prove that the centers of squares built on the given rectangle's sides are the vertices of a square $O_1O_2O_3O_4$.

First, observe that the diagonals of the squares constructed on the adjacent sides of a rectangle lie on the same straight line and are perpendicular to each other.

Indeed, $\angle O_4AB + \angle BAD + \angle O_1AD = 45° + 90° + 45° = 180°$ and

$$\angle O_1DA + \angle ADC + \angle O_2DC = 45° + 90° + 45° = 180°.$$

We also observe that $\angle AO_1D = 90°$ because the diagonals of a square are perpendicular. So, $O_4O_1 \perp O_1O_2$.

Second, since the squares constructed on the opposite sides of the rectangle are congruent, then their diagonals are of the same length.

Therefore,

$$O_1O_2 = O_1D + DO_2 = O_1A + AO_4 = O_1O_4.$$

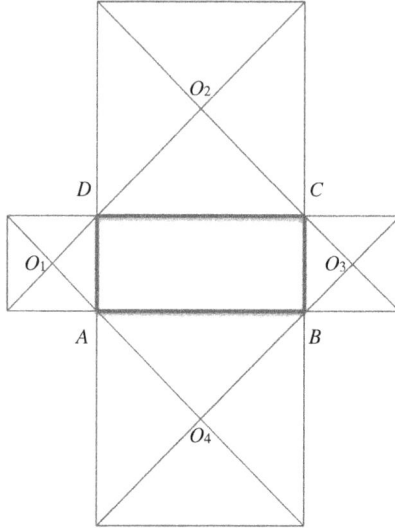

So, we see that the segments $O_1O_4$ and $O_1O_2$ have the same length and are perpendicular to each other.

In a similar way, we can obtain that the other pairs of segments connecting the centers of the squares have the same length and are perpendicular to each other. Therefore, $O_1O_2O_3O_4$ is a square, and our proof is completed.

This problem becomes much more interesting if instead of a rectangle we consider a parallelogram.

**Problem 4.** Prove that the centers of the four squares built on the sides of a parallelogram are the vertices of a square.

**Solution.** $ABCD$ is the given parallelogram. $AMND$, $AFTB$, $BLEC$, and $CKPD$ are the squares built on its sides. $O_1$, $O_2$, $O_3$, and $O_4$ are the centers of the squares, the points of intersections of the diagonals of these squares. We need to prove that $O_1O_2O_3O_4$ is a square.

Dealing with the squares, as we evidenced before, it makes sense to try to apply rotations through a $90°$ angle. Usually, the good results are attained when the center of rotation is selected as the center of a square.

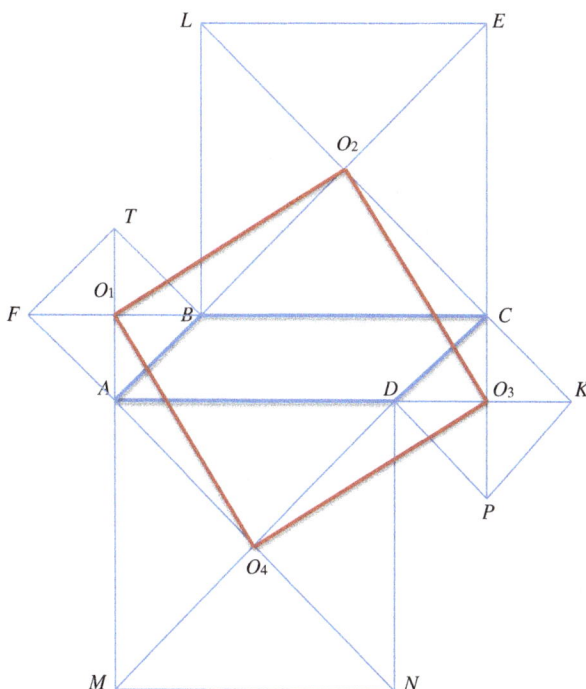

Let's rotate $AFTB$ clockwise through an angle of $90°$ about the center of rotation $O_4$. Since $ADNM$ is a square, its diagonals are congruent and perpendicular to each other. Therefore, $O_4A = O_4D$ and $O_4A \perp O_4D$. It implies that in our rotation $A$ is rotated into $D$.

It is given that $AFTB$ and $DCKP$ are congruent squares (they are built on the opposite sides of the given parallelogram). Observe that $AF = AB = DC$, $AF \perp AB$, and $AB \parallel DC$. Hence, $AF \perp DC$ due to the transitivity rule. It follows that in our rotation $AF$ is rotated into $DC$ and square $AFTB$ is rotated into the congruent square built on $DC$ as on the side. Such a square is $DCKP$. All the elements of $AFTB$ are rotated into the respective elements of $DCKP$. Hence, $O_1$ as the center of $AFTB$ will be rotated into $O_3$, the center of $DCKP$. This implies that by the rotation properties, $O_4O_1 = O_4O_3$ and $O_4O_1 \perp O_4O_3$.

In a similar fashion, if we consider the counterclockwise rotation of $AMND$ through an angle of $90°$ about the center of rotation $O_1$, it is not hard to prove that $O_2O_1 = O_4O_1$ and $O_2O_1 \perp O_4O_1$. There is no need to consider one more similar rotation with the center $O_2$.

We already proved that $O_1O_2O_3O_4$ is a quadrilateral with two pairs of congruent and perpendicular adjacent sides, so it is a square. The problem is solved.

Analyzing this solution, we see that perhaps the toughest task was how to decide what point to pick as the center of rotation and what points and segments to rotate in each specific rotation. The hint here was hidden in the figure itself. Clearly, we were dealing with the pairs of congruent squares $AFTB$ and $DCKP$, $ADNM$ and $BCEL$. Selecting the center of each square as the center of a $90°$ rotation appeared as the most obvious and valid choice. Since a square is rotated into a congruent square, its center is rotated into the center of its image. Applying rotation properties, we arrived at the congruence and perpendicularity of the segments connecting each rotated square's center to the selected invariant center of the rotation, which proved that all four points, the centers of the given squares, are the vertices of the new square.

**Problem 5.** Write the greatest possible number using the digit 2 three times without doing arithmetic operations of addition, subtraction, multiplication, and division.

**Solution.** By using 2 three times, we can create several combinations:

$$222$$
$$2^{2^2} = 16$$
$$22^2 = 484$$
$$2^{22} = 4{,}194{,}304.$$

We see that the greatest result is obtained for $2^{22}$.

What about a similar question about the digit 3? By the same logic, we have

$$333 < 33^3 = 35{,}937 < 3^{3^3} = 3^{27} < 3^{33} = 5{,}559{,}060{,}566{,}555{,}520.$$

Would it be the same outcome for the digit 9?

It's not hard to see that the response will not be similar to the previous two cases.

Clearly, for 9, the greatest number written with three 9s will be $9^{9^9}$ because $9^{9^9} > 9^{99}$.

What about the other digits? Can we make this comparison without the use of a calculator? So, the problem to solve can be stated as follows:

*Can we determine a general rule for writing the greatest possible number using the same digit three times?*

**Solution.** Denote by $m$ some digit $1, \ldots, 9$. When $m = 1$, the greatest possible result is 111. We already saw that for $m = 2$, it is $2^{22}$. Obviously, for all other digits, we need to analyze under what conditions $m^{m^m}$ is greater than $m^{\overline{mm}}$, where the exponent $\overline{mm}$ represents a two-digit number which can be rewritten as $\overline{mm} = 10m + m = 11m$. So, our goal is to compare $m^{m^m}$ and $m^{11m}$. In other words, we have to find the solutions of the inequality $m^{m^m} > m^{11m}$. Recalling the properties of exponential function, we see that since $m > 1$, the last inequality is reduced to solving the inequality $m^m > 11m$. Since $m > 0$, we can divide by $m$ both sides of the last inequality without changing its sign and get the equivalent inequality $m^{m-1} > 11$.

For $a > 1$, the function $f(x) = a^x$ is increasing; therefore, $m^{m-1}$ will be greater than 11 when $m > 3$. So, for 2 and 3, the greatest result is indeed achieved, as we calculated it, when the number is written as $m^{11m}$. For the rest of the digits, when $m > 3$, the greatest possible outcome is attained in the form $m^{m^m}$.

Starting with one interesting problem, we generated a few similar problems, and solved the general case explaining all possible outcomes for all digits.

The readers may wish to investigate further similar problems for getting the greatest possible number written, for instance, using each digit four times.

## 8.3 Extending a Problem

When solving mathematical problems, we often use a technique of recasting objects in one way or another. One of those tricks consists of adding and subtracting the same term in the expression, preserving the total, but making it helpful to rearrange the way you need it. For instance, if you already have in some expression $a^2$ and $b^2$ and your goal is to get to a perfect square, you add and subtract $2ab$. Then, you just group the three terms together to complete a perfect square and work with the rest of the expression depending on the goal you want to achieve in the particular problem. This technique appears to be very helpful in finding new relationships and making a solution manageable. The following old riddle gives a good example of this concept.

**Problem 6.** A father of three sons, understanding he was to die soon, announced his will to his children: "Divide my 17 camels among you in such a proportion: one-half to the eldest son, one-third to the second son, and one-ninth to the youngest son". The following day he passed away, and his sons tried to divide the camels according to their father's will. However, the brothers soon started to argue since they were facing an unsolvable dilemma; 17 is not divisible by 2, 3, and 9. While they were bickering, an old sage was passing by on a camel, and he offered them his help in solving this problem. What did he suggest to fulfill the father's will?

**Solution.** The sage added his own camel to 17 to get 18 camels in total. Then, he gave one half (nine camels) to the older brother, one third (six camels) to the middle brother, and one ninth (two camels) to the younger brother. As each brother got his share of the will, the sage took the remaining camel $(9 + 6 + 2 = 17)$! Since the remaining camel was his own camel, everybody was happy and the problem was solved.

Being a little creative, as we will see in the following Problem 7, one may even get some unexpected benefits by applying the illustrated concept.

**Problem 7.** A government program allows people to collect empty milk bottles and exchange them for full bottles of milk. Four empty bottles may be exchanged for one full bottle. How many bottles of milk can a family drink if they have collected 24 empty bottles?

**Solution.** By exchanging 24 empty bottles for six full bottles of milk, the family can drink all of them, and can get one more bottle full of milk after exchanging four new empty bottles for one bottle full of milk. So, the family can drink $6 + 1 = 7$ bottles of milk and have three remaining empty bottles. Should you stop here and give an answer to the problem as seven full bottles? Yes, you can. But, you can also use some brain power and have a better outcome from your assets. How about borrowing one empty bottle from a friend, for example, and add it to the three empty bottles they already had? It will allow getting one more bottle full of milk in exchange for the four empty bottles. After drinking this "bonus" bottle, the family can return the borrowed empty bottle to the person they got it from. So, now, the family can have a total of eight full bottles of milk and owe nothing to anybody.

**Answer:** Eight bottles full of milk.

Using this technique, you can design lots of interesting problems. Some of them may look like cute problem-jokes, but you may get unexpected and interesting generalizations while analyzing the outcome, as it will be demonstrated in the following Problems 8 and 9.

**Problem-joke 8.** A farmer sold to the first customer half of all his apples and one-half of an apple. He sold one half of the remaining apples and one-half of an apple to the second customer. Then, he sold one half of the remaining apples and one-half of an apple to the third customer, and he repeated this process until he sold the final one half of the remaining apples and one-half of an apple to the seventh customer. After that sale, he had no apples left. How many apples did he originally have?

**Solution.** I imagine the raised eyebrows of confused readers. What a strange way of selling apples? Why did he sell an extra one-half of an apple every time to complete each sales transaction? As we said, this is a problem-joke. Let's assume that's what the customers demanded from him. It is worthwhile to make this assumption and work the problem out, because it leads to interesting observations and thoughts dealing with number series.

Let $x$ be the number of apples a farmer had available for sale. To solve the problem, we will apply a trick similar to the one used in solving the camels' allocation problem. We need to introduce an auxiliary element that will help make the solution manageable. Finalizing the first sale, half of the apples and one half of an apple were sold:

$$\frac{x}{2} + \frac{1}{2} = \frac{x+1}{2}. \qquad (*)$$

To distinguish our auxiliary element and keep it separate, let's assume that he had all the green apples and there was one additional red apple. Going back to the equality (*), we then can state that exactly one half of all the available apples was sold. Therefore, after every next sale, the remaining apples available for sale will be divided in half. After the seventh sale, this number will be reduced in $2^7 = 128$ times. All that is left is the only red apple remaining on hand. So, we arrive at the conclusion that a farmer had 128 apples before the first sale, of which 127 were the green apples and one red apple (an auxiliary element needed to solve the problem, but not really belonging to the farmer).

Using algebraic terms, we can say that the problem is reduced to solving the equation

$$\frac{x+1}{2} + \frac{x+1}{2^2} + \frac{x+1}{2^3} + \cdots + \frac{x+1}{2^7} = x.$$

Factoring out $(x+1)$, we get $(x+1)\left(\frac{1}{2} + \frac{1}{2^2} + \cdots + \frac{1}{2^7}\right) = x$. Observing that we have the sum of the seven terms of the geometric series with the common ratio $\frac{1}{2}$, the equation is further simplified to

$$\frac{x}{x+1} = 1 - \left(\frac{1}{2}\right)^7.$$

Solving this equation gives $x = 127$.

**Answer:** There were 127 apples available for sale.

**Problem 9.** Let's now consider the extension of the previous problem to $k$ customers ($k$ is some natural number).

**Solution.** Letting $x_k$ be the number of green apples available for sale, we can state that after the first sale $x_k - x_{k-1} = \frac{x_k}{2} + \frac{1}{2}$. Multiplying both sides by 2 gives us the recurrent formula for the series $\{x_k\}$:

$$x_k = 2x_{k-1} + 1. \tag{1}$$

By adding the red apple, we essentially made a substitution of variable, leading to the new series $y_n = x_n + 1$. Expressing $x_n = y_n - 1$ and substituting this into (1) gives $y_k - 1 = 2(y_{k-1} - 1) + 1$, which is simplified to $y_k = 2y_{k-1}$. As we see, $\{y_k\}$ represents the geometric sequence with the common ratio 2 and, therefore, we can write its explicit formula for the $n$th term as $y_n = 2^n y_0$ (where $y_0$ is the starting term in the sequence). Going back to the series $\{x_k\}$, it implies that $x_n = 2^n - 1$.

We can now extend our problem further and consider the general case when the series $\{x_k\}$ is defined by the recurrent formula

$$x_k = qx_{k-1} + d. \tag{2}$$

If $q = 1$, then $\{x_k\}$ becomes an arithmetic sequence and its explicit formula is

$$x_n = x_0 + d(n - 1).$$

If $q \neq 1$ and $d = 0$, then $\{x_k\}$ is a geometric sequence and $x_n = q^n x_0$.

Finally, if $q \neq 1$ and $d \neq 0$, we can find $z$ such that the series $y_n = x_n + z$ is a geometric sequence. Substituting $x_k = y_k - z$ and $x_{k-1} = y_{k-1} - z$ into (2) yields $y_k - z = q(y_{k-1} - z) + d$, which simplifies to $y_k = qy_{k-1} + z(1 - q) + d$.

Selecting $z$ in such a way that $z(1 - q) + d = 0$, gives $y_k = qy_{k-1}$, which implies that the explicit formula for $\{y_n\}$ can be written as $y_n = q^n y_0$. So, we determine the formula for the $n$th term for the series $\{x_n\}$ as

$$x_n = q^n(x_0 + z) - z, \text{ where } z = \frac{d}{q - 1}.$$

Now, going back to our Problem 8, we can see that $q = 2$, $x_0 = 0$, $d = 1$, and $z = 1$ (red apple).

Of all regular polygons, the regular pentagon is perhaps the one with the most captivating and thought-provoking properties that lead to a myriad of various, not necessarily related, problems. Let's start with the following construction challenge below.

**Problem 10.** Using classical Euclidian tools of a compass and a straightedge, construct an angle of $72°$.

**Solution.** This problem can be solved in several different ways. We recommend readers explore different approaches and compare them. We will focus here on a technique of entertaining a preliminary problem resembling the given one but simpler to solve. Such a problem should enlighten the steps to take or even give a straight path in our solution.

Observing that $72° = 360° : 5$, the problem is equivalent to dividing the circumference into five equal parts. The central angle formed by connecting the center of a circle with any two of five points equally dividing a circumference will be a $72°$ angle. To rephrase the problem, we need to find a way to construct a segment with the length of a side of a regular pentagon. In order to do this, we will first consider an auxiliary problem. We will solve a similar problem of constructing an angle of $36°$, which is equivalent to dividing a circumference into 10 equal parts.

Assume the problem is solved and we have a regular 10-gon inscribed in the unitary circle with the center $O$ (radius of the circle equals 1). Points $A_1$ and $A_2$ are the adjacent vertices of our regular 10-gon. $\angle A_1 O A_2 = 360° : 10 = 36°$, then in the isosceles triangle $O A_1 A_2 (O A_1 = O A_2$ as radii) each angle by the base $A_1 A_2$ equals $\frac{180° - 36°}{2} = 72°, \angle O A_1 A_2 = \angle O A_2 A_1 = 72°$.

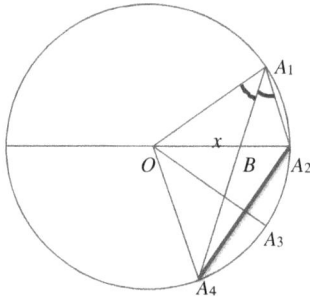

At this point, we can already see the benefits of introducing the preliminary problem of constructing an angle of $36°$. As soon as we construct $A_1 A_2$, we will solve the original problem as well; by connecting $O$ with $A_1$ and $A_2$, we will get the desired angle $\angle O A_1 A_2 = \angle O A_2 A_1 = 72°$. It turned out that unexpectedly we can get the desired result not as predicted at first, through the central angle formed by connecting the center of the circle with the vertices of a regular pentagon, but rather as an angle by the base of an isosceles triangle $O A_1 A_2$.

Draw the bisector $A_1 B$ of the angle $O A_1 A_2$. $\angle O A_1 B = \angle B A_1 A_2 = 72° : 2 = 36°$. Therefore, $O B A_1$ and $B A_1 A_2$ are isosceles triangles ($O B = B A_1 = A_1 A_2$). Since triangles $A_1 O A_2$ and $B A_1 A_2$ have the respective equal angles, they are similar. Let $O B = x$, then $B A_2 = 1 - x$ and from the similarity of the above

triangles, we get

$$\frac{x}{1-x} = \frac{1}{x}, \text{ or equivalently, } x^2 + x - 1 = 0.$$

Solving this quadratic equation gives $x_1 = \frac{-1+\sqrt{5}}{2}$ or $x_2 = \frac{-1-\sqrt{5}}{2}$.

Since $x = OB = BA_1 = A_1A_2$, then the positive root $\frac{-1+\sqrt{5}}{2}$ represents the length of the side of the regular 10-gon. The construction of the segment with length $\frac{-1+\sqrt{5}}{2}$ is easily solvable with a straight-edge and compass. One can construct a right triangle with sides 1 and 2. By the Pythagorean Theorem, its hypotenuse has the length of $\sqrt{5}$. Reducing it by 1 and dividing in half, we get the segment of the desired length.

So, our original problem is reduced to dividing a unitary circle into 10 equal parts and connecting two of those points on a circumference to the center of the circle. The angles by the base of the isosceles triangle $A_1OA_2$ equal 72°. We have reached the desired goal and we can stop at this point, but it worth making a few more interesting observations from our solution.

Making a one step further, we can complete the solution of the constructing of a regular pentagon. Indeed, connecting $A_1$ and $A_3$ or $A_2$ and $A_4$ (where $A_4$ lies on the same circumference and $A_3A_4 = A_3A_2$), we get the side of a regular pentagon.

We suggest that ambitious readers investigate further if the second root of our quadratic equation, $x_2 = \frac{-1-\sqrt{5}}{2}$, has any geometric sense. You should be pleasantly surprised to get a star-shaped decagon; the hint is to locate point $B_1$ to the left of $O$ such that

$$OB_1 = |x_2| = \left|\frac{-1-\sqrt{5}}{2}\right| = \frac{1+\sqrt{5}}{2}.$$

It merits to mention that the found length of the side of the regular 10-gon, the number $\frac{-1+\sqrt{5}}{2}$, is a conjugate to the number $\varphi = \frac{1+\sqrt{5}}{2}$

known as the so-called *golden ratio*, famous for its connection to the Fibonacci numbers and its various applications in many geometrical constructions. An isosceles triangle with a 36° vertex angle and two 72° base angles is called the *golden triangle*; it carries the golden ratio throughout.

By playing around and slightly modifying the achieved results during our solution, we can create many interesting problems.

**Problem 11.** Given points $A_1$ and $A_2$ on the unitary circle (its radius $r = 1$) with the center $O$, such that the central angle $\angle A_1 O A_2 = 36°$, prove that the segment connecting point $A_4$ of the intersection of an angle bisector of the angle $O A_1 A_2$ with the circumference and $A_2$ represents the side of a regular pentagon inscribed in the circle. Find its length.

**Problem 12.** Find the length of a diagonal in a regular pentagon, being given that its side equals 1.

**Problem 13.** A regular decagon of side 1 is inscribed in a circle. Find the radius of the circle.

**Problem 14.** Given a regular pentagon $ABCDE$ of side 1 whose diagonals intersect at points $A'$, $B'$,$C'$, $D'$, and $E'$ (see the following figure), find the ratio of areas of $ABCDE$ and $A'B'C'D'E'$.

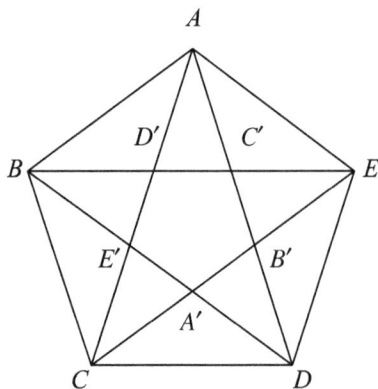

**Problem 15.** Prove that the area of the colored portion of the star (its vertices are the vertices of the regular pentagon) is exactly half the area of the whole star (Offered by N. Avilov, *Quantum*, March/April 1992).

**Problem 16.** The star is formed by drawing the diagonals in a regular pentagon. Is it possible to cut this star into two pieces such that by putting them together we get a centrosymmetric polygon (a polygon that possesses point symmetry)?

**Problem 17.** Can you describe a quadrilateral such that one can move any one of its vertices to some other place and get a new quadrilateral congruent to the original one?

When solving a problem, especially a difficult one, it is very important to be open-minded and to pay attention not just to the short-term goal of finding an answer to the question asked, but to concentrate on possible bypass results that can lead you to pose new questions and hypotheses. After all, as the prominent German mathematician Georg Cantor (1845–1918) stated, "In mathematics the art of proposing a question must be held of higher value than solving it".

While facing the golden ratio in our Problem 10, one may come across its numerous properties and connections to the Fibonacci numbers. This might ignite "inventions" of lots of problems and intriguing questions in number theory.

A good example of the unexpected occurrence of the golden ratio and the Fibonacci numbers helping solving a problem was demonstrated in Chapter 11 of *The Equations World*, Dover Publications, 2019. We considered the following interesting problem:

There is the given equation $4x^k + (x+1)^2 = y^2$. Prove that

(1) There are no natural solutions when $k = 1$.
(2) There are at least two natural solutions when $k = 2$.
(3) There exist infinite number of natural solutions when $k = 2$.

While proving part 3 that the Diophantine equation has infinitely many natural solutions, in the development of our solution, we arrived at the sequence of numbers, which turned out to be the subset of the Fibonacci numbers, all its odd terms. We established an interesting relationship between the odd Fibonacci numbers $F_i$ and numbers $a_i = 5F_i^2 - 4$, and concluded that for each number $F_i$ the respective $a_i$ is a perfect square. This exposed a question for a similar property for the even Fibonacci numbers and numbers $a_i = 5F_i^2 + 4$, i.e., whether for each number $F_i$ the respective $a_i$, expressed now as $a_i = 5F_i^2 + 4$, is a perfect square. Finding the positive answer to the posed question (absolutely unexpectedly, it has nothing to do with the original problem we were solving!), we arrived at formulating necessary and sufficient conditions for identifying a Fibonacci number (a term in the Fibonacci sequence):

> *A positive integer $F_n$ is a Fibonacci number if and only if $5F_n{}^2 + 4$ is a perfect square for even number $n$ or $5F_n{}^2 - 4$ is a perfect square for odd number $n$.*

So, while solving the Diophantine equation, we arrived at some bypassing results, and posed several questions, inventing new problems. After we managed to find the answers to those questions, our results led us to formulating a general statement for identifying the Fibonacci series numbers, which is a very interesting outcome on its own. We omit discussing the proofs and solutions here. Ambitious readers can find the details in *The Equations World* or it should be a good exercise to seek their own solutions for the above problems, and see if some other interesting variations or generalizations can be derived.

During our solution of Problem 10, we recognized that the problem was reduced to constructing a regular decagon inscribed in a circle. Conceiving the plan for a solution, we came across setting up and solving a quadratic equation, the positive solution of which represented the length of the side of the regular decagon, an interesting bypass result on its own. Since it was expressed as $\frac{-1+\sqrt{5}}{2}$, then the segment with such a length is straightedge-and-compass constructible, and we concluded that the regular decagon is constructible with two classic construction tools. Furthermore, applying this outcome allowed us to perform the regular pentagon construction as well. Should it be natural to ponder a general related inquiry under what conditions a regular $n$-gon construction is possible with a straightedge and compass?

A classic example of coming across new interesting relationships and new problems not related to the originally considered issues is presented by the great German mathematician Carl Friedrich Gauss' (1777–1855) discoveries during his solution of the regular 17-gon construction. While he was considering a pure geometrical problem and raised a question regarding the conditions for the constructability of a regular $n$-gon, he came across cyclotomic equations and the issues related to their solvability in radicals. Gauss proved sufficient conditions for the constructability of regular polygons. The proof of the necessary conditions was given by the French mathematician Pierre Laurent Wantzel (1814–1848) completing the now well-known Gauss-Wantzel Theorem:

> *A regular n-gon can be constructed with the compass and straightedge if and only if n, the number of its sides, is the product of a power of 2 and any number of distinct Fermat primes: $n = 2^k \cdot p_1 \cdot p_2 \cdots \cdot p_m$, (k and m are natural numbers; $p_i$ are distinct Fermat primes), where Fermat prime numbers have the form $2^{2^n} + 1$.*

Carl Gauss

What an amazing connection between geometrical and algebraic issues! Ambitious readers can find out more about this fascinating topic in mathematical literature; for instance, it was mentioned and discussed in more detail in Chapter 6 of *The Equations World*, Dover Publications, 2019.

Another classic example of getting unexpected bypass results not related to the originally considered problem is presented by Leonhard Euler's (1707–1783) discovery of a line connecting the orthocenter, centroid, and circumcenter of a triangle, now known as *Euler's Line.* Euler was solving the following problem:

*Given four points, circumcenter, centroid, orthocenter, and incenter of a triangle, reconstruct the triangle.*

Leonhard Euler

The results he incidentally obtained during the solving process revealed the remarkable fact that the three classic centers of a triangle are collinear, and the centroid lies between the circumcenter and orthocenter and is twice as far from the orthocenter as it is from the circumcenter. It merits pointing out that we also incidentally got the same result over applying a homothety twice while solving Problem 11 from Chapter 6. Our goal was to prove the existence of the orthocenter of a triangle, but in fact, we did not just prove the required statement. We got a much more interesting outcome, after observing and analyzing the facts obtained during the solution process.

During the Second World War, the prominent English mathematician Alan Turing (1912–1954) was working on breaking German ciphers. What started as a pure mathematical problem in cryptography led to the creation of the Turing machine, the first model of a general-purpose computer.

Alan Turing

This is an amazing example of the influence the solving of mathematical problems has on human life. While he was creating algorithms for solving the problem of deciphering the German Enigma machine, Alan Turing posed new questions and problems, which ultimately laid out the foundations for theoretical computer science and artificial intelligence. It has been estimated that his successful solution of the original problem of breaking German ciphers shortened the war in Europe by more than 2 years and saved over 14 million lives. It is truly an astonishing achievement. The bypassing results of his work though, which led to inventing and developing computer science, had an absolutely immeasurable influence on human life!

# Chapter 9

# Related Problems

## 9.1 Problems Relying on the Same Property

To become a successful problem solver, it is crucial to be able to identify the problem's most important conditions and, if possible, associate the problem to a related example that can be solved through using similar techniques and relying on the same property or theorem. By recognizing a problem as belonging to a certain family of similar problems, you should be able to contemplate your plan for a solution in a very specific direction. By properly pointing out the most critical given attribute, you utilize it to establish links among other elements to find the path to the desired outcome. Furthermore, while identifying problems-siblings and studying their properties, you may come across lots of new interesting problems and even generalizations, hence we will consider this chapter as a continuation of our discussion of new inventions and discoveries that we started in the previous chapter.

Let's consider an important geometrical property that can be used for solving many interesting problems, which rely on this property:

> *The median to the hypotenuse of a right triangle equals one-half of the hypotenuse.*

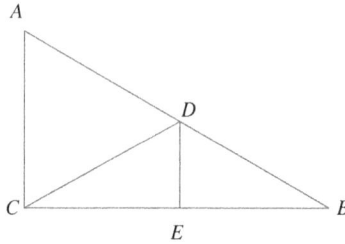

**Proof.**   Let us consider the right triangle $ACB$ with the right angle $C$ and let $CD$ be the median drawn from the vertex $C$ to the hypotenuse $AB$. We need to prove that the length of the median $CD$ is half the length of $AB$.                                    □

Draw the straight line $DE$ passing through the midpoint $D$ parallel to the leg $AC$ till its intersection with $CB$ at $E$. It is given that $ACB$ is the right angle. The angles $BED$ and $BCA$ are congruent as they are corresponding angles at the parallel lines $AC$ and $DE$ and the transversal $BC$. Therefore, the angle $BED$ is the right angle.

Now, since the straight line $DE$ passes through the midpoint $D$ and is parallel to $AC$, it cuts the side $CB$ in two congruent segments of equal length: $CE = EB$. Triangles $CED$ and $BED$ are right triangles that have congruent legs $CE$ and $BE$ and the common leg $DE$. Hence, these triangles are congruent by side-angle-side (SAS). It implies that the segments $CD$ and $DB$ are congruent as the corresponding sides of these triangles. Since $DB$ has a length that is half the length of the hypotenuse $AB$, we have proved that the median $CD$ has a length that is half the length of the hypotenuse.

The converse statement holds as well, that is,

> *if in a triangle the median equals one-half of the side to which it is dropped, then the triangle is a right triangle with the right angle at the vertex from which the median was dropped.*

We leave the proof of this statement for the readers.

Assessing a problem often starts with the question — can you restate a problem? By looking at the same problem from a different angle, we can get a deeper understanding of its particulars and by applying the same property in several scenarios make the solution manageable. For instance, the considered properties can be restated as follows:

*The midpoint of a hypotenuse in a right triangle is the center of its circumscribed circle*

or

*if A, B, and C are distinct points on a circle where AC is a diameter, then the angle ABC is a right angle* (this statement is known as Thales's theorem)

or

*an angle inscribed in a semicircle is a right angle.*

Depending on the problem you are solving, you may want to use one of the above equivalent statements.

**Problem 1.** Construct a triangle $ABC$, given its vertices $A$ and $B$, and its orthocenter $H$.

**Solution.**

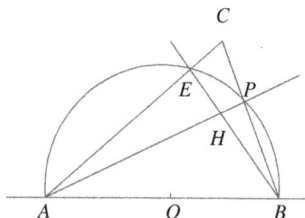

This construction calls for a triangle $ABC$, given its vertices $A$ and $B$ and $H$, the point of intersection of its altitudes. The straight lines passing through $A$ and $H$ and through $B$ and $H$ contain the triangle's altitudes. Using the property that any angle inscribed in a circle and subtended by a diameter is a right angle (another way of expressing the same idea that any angle inscribed in a semicircle is a right angle), we obtain that these straight lines will intersect the semicircle with diameter $AB$ at points $E$ and $P$, such that $\angle AEB = \angle APB = 90°$. By extending $AE$ and $BP$ till their intersection, we can get the third vertex $C$ of the triangle $ABC$. This concludes our analysis. The construction steps are as follows:

Having points $A$ and $B$, we can find the midpoint of $AB$ and draw a circle with the center at the midpoint of $AB$ and with the radius that equals $\frac{1}{2}AB$. $AH$ and $BH$ intersect this circle at $P$ and $E$, respectively. Draw $BP$ and $AE$ till their intersection at $C$, the third vertex of $\triangle ABC$. We arrive at the desired triangle $ABC$.

**Problem 2.** Given a rectangle inscribed in a circle, prove that the sum of the squares of the distances from any point on a circle to each vertex of a rectangle is a constant number.

**Solution.**

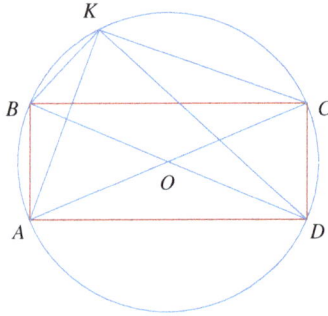

$ABCD$ is the rectangle inscribed in the circle with the center $O$, the point of intersection of its diagonals. Our goal is to prove that for any point $K$ located on this circle, the following equality holds true: $AK^2 + BK^2 + CK^2 + DK^2 = c$, where $c$ is some constant number. $BD$ and $AC$ are the diameters of the circle, $K$ lies on the circle. Therefore, $AKC$ and $BKD$ are the right triangles with right angles at vertex $K$. Once again, we relied on the property that an angle inscribed in a semicircle is a right angle. Denoting the diameter of the circle by $d$ and applying the Pythagorean Theorem to each of the right triangles $AKC$ and $BKD$ gives

$$AK^2 + CK^2 = d^2,$$
$$BK^2 + DK^2 = d^2.$$

Adding these two equalities, we obtain $AK^2 + BK^2 + CK^2 + DK^2 = 2d^2$. Since $d$ is a fixed constant number for the given circle, we arrive at the desired result.

**Problem 3.** Given the right angle with vertex $O$ and segment $AB$ with end points located on the sides of the angle, the segment moves in such a way that its end points slide on the sides of the given angle and its length stays fixed. Find the locus of the centroids of all the triangles with the vertices at $A$, $B$, and $O$.

**Solution.**

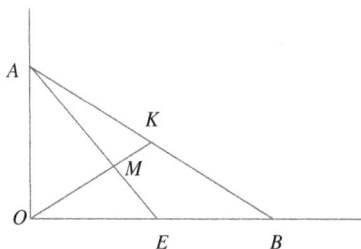

Consider triangle $AOB$ and its medians $OK$ and $AE$. Let $M$ be the point of intersection of $OK$ and $AE$. Since the medians of a triangle meet at a point that is two-thirds of their lengths from the vertices, then $OM = \frac{2}{3}OK$. Now, observing that $AOB$ is a right triangle, we have $OK = AK = KB = \frac{1}{2} AB$ ($K$ is the center of the circle circumscribed over $AOB$. So, $OK$, $AK$, and $KB$ are the radii of that circle).

Therefore, for any location of point $K$ for the moving $AB$, the distance from $O$ to $M$ will be the same and equal $OM = \frac{2}{3}OK = \frac{2}{3} \cdot \frac{1}{2}AB = \frac{1}{3}AB$. The locus of all centroids of the triangles formed by moving $AB$ and fixed point $O$ will be $\frac{1}{4}$ of the circle with the center at $O$ and radius $r = \frac{1}{3}AB$.

**Problem 4.** Given right triangle $ACB$ ($\angle ACB = 90°$), in which the hypotenuse is four times of the length of the altitude dropped to it, find the acute angles of $\triangle ACB$.

**Solution.**

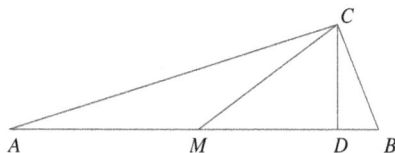

It is given that $\angle ACB = 90°$, $CD \perp AB$, and $AB = 4CD$. We have to find the angle measures for $CAB$ and $CBA$.

Let's draw median $CM$ to $AB$. Hence, $CM = AM = MB = \frac{1}{2}AB$. In the right triangle $MDC$, we see that the length of leg $CD$ is half of the length of its hypotenuse $CM$. Indeed, $CD = \frac{1}{4}AB = \frac{1}{2}CM$.

It follows that $\angle CMD = 30°$. Angles $AMC$ and $CMD$ are supplementary, so $\angle AMC = 180° - 30° = 150°$. Once again, referring to the equality of $AM$ and $MC$, we see that triangle $AMC$ is isosceles and, respectively,

$$\angle MAC = \angle MCA = \frac{180° - 150°}{2} = 15°.$$

Finally, getting back to the right triangle $ACB$, we have $\angle CBA = 90° - 15° = 75°$.

Answer: $\angle CAB = 15°, \angle CBA = 75°$.

**Problem 5.** Given a segment $AB$, find the locus of vertices $C$ of all acute-angled triangles $ABC$.

**Solution.** The problem may look extremely difficult if we would not have referred to the basic problem we considered above and stated as *An angle inscribed in a semicircle is a right angle.* All points $C$ selected on the circumference of the circle with the diameter $AB$ will be the vertices of right triangles (with angle $C$ equal to $90°$); points located inside this circle will be the vertices of obtuse triangles $ABC$ (with obtuse angle at vertex $C$); points located outside the two lines $m$ and $n$ tangent to the circle and passing through $A$ and $B$ will be the vertices of obtuse triangles (with obtuse angles by vertices $A$ and $B$); points located on two tangent lines $m$ and $n$ will be the vertices of right triangles (with angles at $A$ and $B$ equal to $90°$). Finally, all the points outside of the circle with the diameter $AB$ and inside the two parallel tangents $m$ and $n$ to this circle passing through $A$ and $B$ will represent the locus of vertices $C$ of all acute-angled triangles $ABC$.

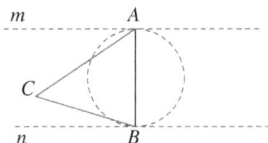

We invite the readers to investigate the rigid proofs of the above observations.

As we mentioned at the beginning of the chapter, one of the very effective techniques in problem solving is restating a problem. The solutions in all the problems-relatives above had relied on the same property. However, we would pick every time from one of this property's four different (yet still equivalent!) statements the one that was the most appropriate for the considered problem. And, you should be able to clearly see now what a great impact these selections had made in finding the right path to a solution. It merits paying more close attention and having a detailed discussion of this important problem-solving aspect, which we are going to explore next.

## 9.2   Restating a Problem

Besides the examples studied at the beginning of this chapter, while solving several problems in our book, we restated them in order to look at the equivalent problems, and clarify the path for their solutions. A good example was given in Chapter 8 when we restated Problem 1.4 and completely changed the approach to its solution from the original perspective. Instead of solving a problem of inscribing $\triangle MNP$ into $\triangle ABC$, we solved a problem of circumscribing of $\triangle ABC$ over $\triangle MNP$. It was the same problem, merely stated differently. But, what a difference it made in terms of contemplating a plan for its solution!

This chapter is devoted to problems-relatives. Can we claim that the restated problem is a relative to the original one? No, it is exactly the same problem. We have to clearly understand that the problem after restatement should not change at all. We are still solving the same problem, while trying to clarify and better visualize the given information. However, by doing this, we are usually not just looking for the ways to make a problem's solution manageable but open a myriad of opportunities to find many new relatives or siblings of the original problem, i.e., making new "inventions" as well.

Restating a problem allows us to look at a problem from a different angle, which sometimes leads us to completely change the direction to our solution process. The simplest instance of restating a problem is translating a word problem into algebraic terms by setting up the equations or inequalities connecting the given conditions.

**Problem 6.** Elly and Liana went to a bookstore to buy the last *Harry Potter* book. As they got to the store, Elly realized that she was short of $13 to buy the book, and Liana figured that she was short of $2 to buy the book. As they counted their combined funds, it was a disappointing outcome — they still were short of money to buy even a single book. How much did the book cost (assume the cost is expressed in dollars with no cents, i.e., it is an integer number, and each girl has some money on hand)?

This is not a complicated problem at all. But at the first glance it may appear that we are not given enough information to find a definite answer to the question asked. We perhaps have this feeling because of the problem's wording. It is usually more difficult to envision the given attributes and relate them together when reading a problem in a "denial" statement. Indeed, each girl lacked the funds to buy a book. By combining their money, they still didn't have enough to even buy one, not two, book. So, by reading this problem, we are getting a negative perception of the events, which affects the way we think about the problem.

Let's try to slightly restate the problem and look at it from a different angle:

> *The cost of a Harry Potter book is $13 more than what Elly has on hand and $2 more than what Liana has on hand. What is the cost of the book if we know that by combining their funds, the girls still can't afford to buy even a single book?*

Now, it is much easier to establish some links between the given conditions and translate the problem into algebra terms. Assume the cost of the book is $x$ dollars (which must be, of course, a positive integer). Then, Elly has $(x-13)$ dollars, and Liana has $(x-2)$ dollars. Adding these two amounts, they still get some number that is less than $x$, the cost of the book, that is, $(x - 13) + (x - 2) < x$.

Simplifying the last inequality, we get $2x - 15 < x$, from which $x < 15$. So, we obtained that the cost of the book is less than $15. Recall now that Elly has $(x - 13)$ dollars on hand. We know that $x < 15$ and, clearly, $(x - 13)$ is some integer. Thus, the only possible integer value $x$ can attain is $14. Therefore, the cost of the book is $14. Elly has just $1 and Liana has $12, that's why they cannot afford to buy the book.

Most likely, one can get to the same conclusion by carefully analyzing the data of the original problem. But, it appears that the restated problem became easier to comprehend. Our restatement also helped to make a straightforward transition into algebra terms.

In many cases, we restate a problem in words, and it is very important to do this properly and keep all the given elements and attributes intact. Many game and combinatorial problems present good examples of doing this, as it is demonstrated in the following chess etude Problem 7.

**Problem 7.** (The Knight's tour problem) Is it possible to make a chess knight move on every square out of the 64 total squares of a checkered board by visiting each square only once?

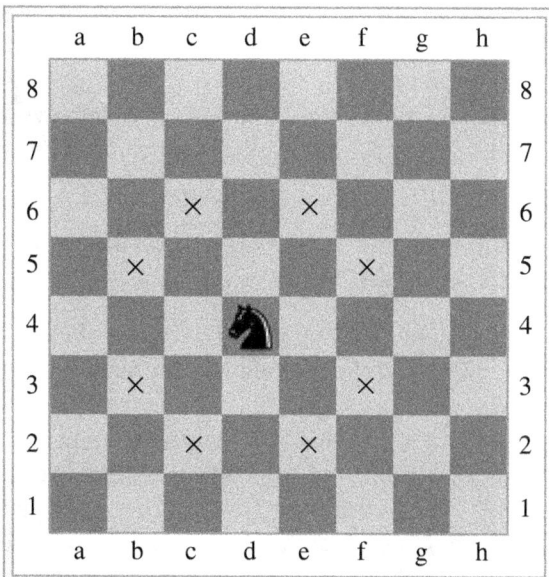

It is one of those classic problems solvable with the use of graph theory; it is asking for a Hamiltonian path in a certain graph.

This problem becomes more transparent if we restate it in the following way:

> *All the boxes except one of a 64-square checkboard are occupied by the chess knights. Is it possible to make exactly one move by each knight?*

We leave this interesting question for our readers who are chess game fans to resolve (one of the possible solutions is presented in the "Solutions to Problems" section).

Meanwhile, we are going to use the game of chess one more time to demonstrate its surprising application in the following Problem 8. By selecting proper language to restate a problem, sometimes we manage to find a problem-sibling that is much easier to solve. This may enlighten the solution of the original problem in an unexpected and elegant way.

**Problem 8.** In how many ways can you assign $n$ workers to do $n$ different jobs so each job is performed by only one person?

Even after reading this problem several times, it still looks confusing and hardly accessible. What is it about? What is the best approach to the problem?

We suggest that instead of solving this problem, we consider an analogy of a chess game problem that would clarify the essence of the original problem.

**Problem 9.** In how many ways can you locate $n$ non-interacting chess rooks on an $n \times n$ size checkered board?

For those readers not familiar with the game of chess, a rook can move on any number of squares along a rank or file, but cannot leap over other pieces.

Let's now establish the following correspondence: the horizontal lines of a checkboard represent the workers and the vertical lines represent the jobs assigned among them. When a worker $i$ gets the assignment for a job $j$, we place a rook at the square located at the intersection of lines $i$ and $j$.

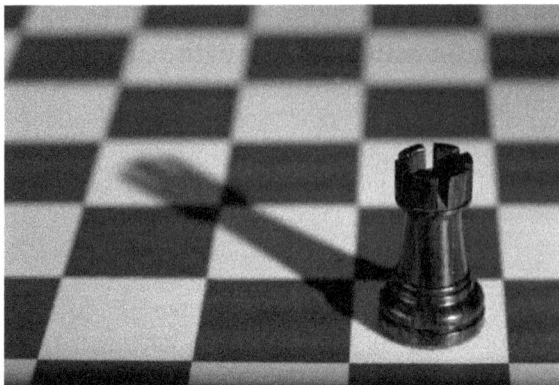

Since every worker has to be assigned to only one job, and each job has to be performed by only one worker, then as a result of positioning the rooks on a checkboard, all $n$ rooks will be located on the intersections of just one horizontal and one vertical line, so they cannot "threaten" each other. By using chess game rules, we managed to show a vivid and clear interpretation of our original financial problem. There are $n$ possibilities of placing the first rook on a checkered board. Once it's done, there are $n-1$ possibilities for the next rook, and so no. One gets the answer $n! = 1 \cdot 2 \cdot \ldots \cdot n$ pretty easily.

## 9.3 Converse Problems

Studying problems-relatives should be closely connected to discussion of converse problems.

By exchanging the hypothesis and conclusion of a problem, you can come up with a converse problem. Such a problem may present a nice challenge and lead to posing lots of new questions, paving the way to interesting conclusions and even new discoveries.

Students rarely think about the necessity to prove the converse statement of the well-known property or theorem. They sometimes confuse the direct with the converse statement and assume that the validity of the direct statement presumes the validity of the converse statement. But, it is not necessarily true. It is important to understand that the converse always requires an independent proof, even for seemingly "obvious" statements. Let's now go over some examples, starting with the two classic properties-theorems.

**Problem 10.**

*The direct statement*: if the discriminant of a quadratic equation equals 0, then the equation has only one root (two coinciding roots).

*The converse statement*: Prove that if the quadratic equation has only one root, then its discriminant must equal to 0.

**Proof.** Consider the quadratic equation $ax^2 + bx + c = 0 \, (a \neq 0)$ which has only one root. We need to prove that its discriminant must equal 0, i.e., $D = b^2 - 4ac = 0$. $\qquad\square$

We know that by Viète's Formulas for the roots of a quadratic equation,

$$x_1 + x_2 = -\frac{b}{a},$$

$$x_1 \cdot x_2 = \frac{c}{a}.$$

Substituting $x_1 = x_2$ into each of the above formulas gives

$$2x_1 = -\frac{b}{a},$$

$$x_1^2 = \frac{c}{a}.$$

Dividing both sides of the first equality by 2 and then squaring both sides gives $x_1^2 = \frac{b^2}{4a^2}$. Substituting this into the second equality, we have $\frac{b^2}{4a^2} = \frac{c}{a}$. It follows that $b^2 - 4ac = 0$, which completes our proof that $D = b^2 - 4ac = 0$.

While solving Problem 10 in Chapter 3, we referred to the converse of the Pythagorean Theorem. We did not mention though the statement of that theorem. Let's remedy this omission here and consider it as a separate problem to solve.

The Pythagorean Theorem states that in a right triangle the sum of the squares of the legs equals the square of the third side, a hypotenuse.

**Problem 11.** The converse of the Pythagorean Theorem:

> *In a triangle, if the square of one side is equal to the sum of the squares of the other two sides, then the angle opposite to the first side is a right angle.*

The alternative statement can be formulated as

> *For any three positive numbers a, b, and c such that $a^2 + b^2 = c^2$, there exists a triangle with sides a, b, and c, and every such triangle has a right angle between the sides of lengths a and b.*

**Proof.**     There are various proofs of this remarkable theorem. One of the easy choices is to use the *Law of cosines*. It states that for any triangle with sides $a$, $b$, and $c$, and angle $\gamma$ between sides $a$ and $b$, the following equality holds true:

$$a^2 + b^2 - 2ab \cdot \cos\gamma = c^2. \qquad\qquad \square$$

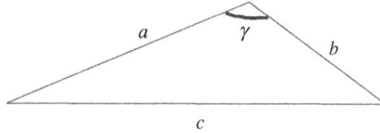

It is given that in our triangle $a^2 + b^2 = c^2$. The *Law of cosines* equality then holds only when $2ab \cdot \cos\gamma = 0$. Since $a \neq 0$ and $b \neq 0$, it follows that $\cos\gamma = 0$. This implies that the angle between $a$ and $b$ has to be a right angle, i.e., $\gamma = 90°$, which is the desired result. The Pythagorean Theorem is one of the most famous in mathematics. Finding different proofs of the Pythagorean Theorem and the converse statement, looking for analogies and generalizations, and studying consequences and uses of them present an ample source of invaluable problem-solving experiences and enjoyment. We will be referring to it one more time at the end of this chapter while talking about generalizations. This classic historical problem and its converse may become an inspiration for a separate mathematical journey for an ambitious reader.

Let's now turn again to Jacob Steiner's problem that we have used in a few previous chapters about the collinearity of the midpoints of the bases of a trapezoid, the point of intersection of its diagonals, and the point of intersection of non-parallel sides (it has been so useful for us throughout the book!), and formulate the converse of it, which appears to be a non-trivial problem to solve.

*The direct statement:* In a trapezoid, the midpoints of the bases are collinear with the point of intersection of the diagonals, and also with the point of intersection of the two non-parallel sides.

**Problem 12.** The converse statement:

> *If in a convex quadrilateral the midpoints of the two sides are collinear with the point of intersection of the diagonals and the point of intersection of the extended two other sides, then this quadrilateral is a trapezoid.*

**Solution.** In the given convex quadrilateral $ABCD$, the diagonals intersect at $O$, the extended sides $AB$ and $DC$ intersect at $E$, and

$N$ and $M$ are the midpoints of $BC$ and $AD$, respectively. It is given that points $E$, $N$, $O$, and $M$ are collinear. We want to prove that $BC \parallel AD$.

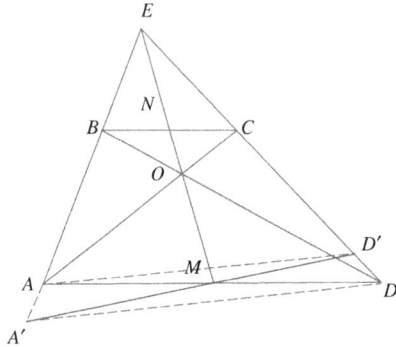

For the proof, we will consider a homothety with the center $E$ and coefficient $k$ such that the midpoint of $BC$, point $N$, is transferred into the midpoint $M$ of $AD$. If we manage to prove that this homothety takes $BC$ into $AD$, then by homothetic properties, $BC \parallel AD$, and we will be done. Assume now that in this homothety the image of $B$ is $A'$ $(A' \neq A)$ and the image of $C$ is $D'$ $(D' \neq D)$. Then, the image of $BC$ is $A'D' \neq AD$. By the definition of a homothety, points $A'$ and $D'$ must be located on the rays $EA$ and $ED$, respectively. Without loss of generality, we assume that $A'$ is located on the extension of the segment $EA$ (in the assumption that $A'$ belongs to the segment $EA$, the rest of the proof will be the same). Since $A'D'$ is the image of $BC$ and $M$ is the image of the midpoint $N$ of $BC$, then $M$ must be the midpoint of $A'D'$. Note that it is given that $M$ is the midpoint of $AD$, and we proved that in our homothety $M$ is the midpoint of $A'D'$. Connecting now points $A$ and $D'$, $A'$ and $D$, we can produce a parallelogram $A'AD'D$, thereby establishing a convex quadrilateral the diagonals of which $AD$ and $A'D'$ bisect each other at $M$. But, this is impossible because the straight lines $A'A$ and $DD'$ are not parallel, they intersect at $E$. We arrived at a contradiction. It allows us to conclude that $A'$ coincides with $A$ and $D'$ coincides with $D$. So, the introduced homothety takes $BC$ into $AD$. Hence, according to the properties of homothetic transformations, these two segments are parallel, $BC \parallel AD$; and indeed, $ABCD$ is a trapezoid, as it was required to be proved.

It is interesting to note that the fact that $O$ lies on $NM$ was never used. So, in our proof, we actually proved that if $E$, $N$, and $M$ are collinear, then $AD$ is parallel to $BC$.

The considered proof is not unique. We strongly suggest that readers search for alternative solutions. Speaking about our approach here, we want to emphasize that applying the properties of homothetic transformations allowed us to get a non-trivial and elegant solution. Why did we decide to apply homethety in our solution? As it was mentioned in Chapter 6, homothetic properties prove to be useful in dealing with collinearity of points. In this case, the most important of the given conditions was that four points $E$, $N$, $O$, and $M$ are collinear. We know that homothetic transformations preserve the ratios of the lengths of the respective segments, and the respective segments are parallel. So, it was a good idea to consider a specific homothety that transfers the midpoint of one of the sides of the quadrilateral into the midpoint of the other side. Our goal was to prove that in such homothety the considered sides are the images of each other, and therefore, must be parallel. It was a successful attempt, and we did get the desired result.

In many cases, the converse statements are harder to prove than their direct versions. The formulated converse statement would not always become a correct statement. You then would need to investigate and decide which additional conditions and restrictions to apply in order for the converse statement to be valid.

For instance, obviously, the following statement is correct: *In any square the diagonals are congruent and perpendicular to each other.*

The converse can be formulated as follows: *If the diagonals of a convex quadrilateral are congruent and perpendicular to each other, then it is a square.* Is this a correct statement? It suffices to show at least one example of a convex quadrilateral satisfying the above conditions, which however, is not a square, to demonstrate that the statement does not hold.

$AC = BD, AC \perp BD$

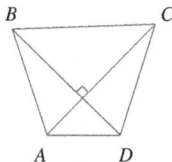

The diagonals of $ABCD$ are perpendicular to each other and have the same length, but $ABCD$ is not a square. So, the converse statement is false.

Speaking about converse problems, we ought to mention the so-called inversion problems. The way of thinking in reverse, i.e., the way of thinking in which you consider the opposite of what you want to achieve, is known as inversion. It is a very powerful thinking tool because it helps in identifying obstacles and errors in problem-solving that are not obvious at the first glance. Applying the inversion thinking technique, instead of asking how to do something, we ask how to not do it; or in other words, instead of thinking forward, we flip it in reverse and think backwards. German mathematician Carl Gustav Jacob Jacobi (1804–1851) is regarded as the father of the "inversion technique" in mathematics. Jacobi was the first Jewish mathematician to be appointed a professor at a German university, at Königsberg University. He made significant contributions in mathematics, among which the most noticeable are developing the theory of ecliptic functions and their relation to the ecliptic theta function, the study of differential equations, and rational mechanics. While teaching his students to solve difficult problems, he encouraged them to utilize the strategy of *man muss immer umkehren*, translated as "invert, always invert". In his belief, this strategy was not just very efficient in solving difficult problems, but it could open up new horizons for mathematical research as well. When applying this technique, the critical question to ask is what would happen if the opposite was true. This concept is well demonstrated in the following Problem 13.

**Problem 13.** The participants of the final stage in horse racing were befuddled by the announcement of the racing rules: the winner of this race would become the last horse at the finish line. What is the best strategy for the participating jockeys to win this race?

**Solution.** First of all, it is important to concentrate on the specifics of the rules for this horse race — it is clearly stated that the winner would be announced the last horse (not a jockey!) at the finish line. This detail is critical in making the decision regarding the winning strategy. Using the inversion technique, the key to solving the problem is to switch all the horses among participating jockeys. This immediately changes the strategy of all jockeys and put them in normal competition mode. Since now each of them will be riding on somebody else's horse, every jockey would try his best to get first to the finish line leaving his own horse behind.

While solving this problem, we were facing the inversion not once but twice. First, the organizers of the race applied an inversion in making the decision for announcing the last horse at the finish line to become a winner (they changed the usual "get to the finish first" to "get to the finish last"). Second, the inversion technique was applied in the decision to switch the horses among jockeys.

Recall now Problem 7 about the chess knight movement across the checkered board by visiting each square only once. In fact, by restating the problem, we utilized the inversion. Indeed, we "substituted" the empty squares on a checkered board in the original problem for the squares occupied by the chess knights in the restated problem. Respectively, the trajectory-route of a chess knight was "substituted" for the route-movement of a single empty checkered board square.

**Problem 14.** A dying old man wants to leave an inheritance to his two sons, his entire land property. Since his only wish is to treat them fairly, he wants to divide his land equally between them. The problem is that the land property is significantly irregular in shape and thus it is hard to cut it in perfectly equal halves. What is the possible strategy to divide the land in a manner that both of his sons are satisfied?

**Solution.** He would need to ask one of his sons to do the measurements and divide the land into equal two parts and to tell him that the other son will have the choice of selecting either of two parts. In this manner, both of them should be happy and feel fairly treated.

This problem is a good example of inverse thinking. To achieve the desired goal, the first son would need to concentrate not on *gaining* some reward, but rather on how *not to fail* in cutting the property into unequal parts, so the other brother would not be able to get the greater piece.

The demonstrated thinking technique is crucial for solving many problems in real life and for making good decisions consistently. Inversion is an essential skill for living a rational life. It has been used by outstanding scientists, philosophers, and innovators throughout history. One of the great examples is given by Charlie Munger, the billionaire investor and business partner of Warren Buffett. He proclaimed that thinking about planning for the opposite of what you want to accomplish gives you a competitive advantage in anything you do. Warren Buffett agreed to this and also advised that it is critical to concentrate on how to not lose money instead of how to make more money. Indeed, we know lots of examples of how some businessmen were successful in *making money*, but they thoughtlessly overspent their funds and ended up in bankruptcy. It is important to think about achieving success, but it is even more important to focus on the inverse instead — how to avoid failure! The powerful benefit of using the inversion technique is that it helps to avoid bad decisions that prevent us from achieving our goals. By thinking about what you need to avoid doing, you can plan ahead to prevent possible failure.

Turning back to mathematics, examining a converse problem to the given problem and applying the inverse technique in problem-solving present great mental exercises which also can be referred to as "inventing of new problems". Considering "conversion game", you can play it, discovering lots of new interesting challenges. Developing the ability of thinking forward and backward and utilizing inversion are very important in benefitting us in solving unconventional problems and making new inventions as well.

## 9.4   Making a "Prognosis". Generalizations

The most useful and effective approach in problem-solving is the combination of analysis with synthesis when we are interchangeably thinking in both directions, from the given data to the question asked (synthesis), and in the opposite direction, from the question asked to the given attributes (analysis). During this process, it is important to be able to make a "prognosis" in justifying the suggested steps to take. While identifying a problem as one that belongs to a specific family, we apply a technique that is usually effective for that specific family, thus making a prognosis about the possible results that lead us in a suggested direction. On the other hand, when facing an unrecognizable problem, our goal is to try to get an enlightening idea for the proper path to take in a problem's solution. A good idea is always subtended by a proper prognosis, or "prediction" of reasons for benefitting from using it. Prognosis is closely connected with intuition and the ability to think creatively.

For a good problem solver, sometimes, many questions remain open even after finalizing the solution of an interesting problem. It pays off to look back and analyze the solution process, especially when some of your predictions did not work out well. Searching for the answers to the following questions might get us unexpected and interesting results:

Can we change a problem's conditions by removing certain restrictions, and look for a general case? Can we extend a problem?

**Problem 15.** Prove that in a convex quadrilateral the sum of the squares of its diagonals equals the doubled sum of the squares of its middle lines (segments connecting the midpoints of the adjacent sides).

**Solution.** In a convex quadrilateral $ABCD$, points $M$, $K$, $N$, and $P$ are the midpoints of $AB$, $BC$, $CD$, and $DA$, respectively. The goal is to prove that

$$AC^2 + BD^2 = 2(MK^2 + KN^2 + NP^2 + PM^2).$$

In order to decide what the best approach to the problem's solution is, we may consider several ideas and make a prognosis about selecting the proper one.

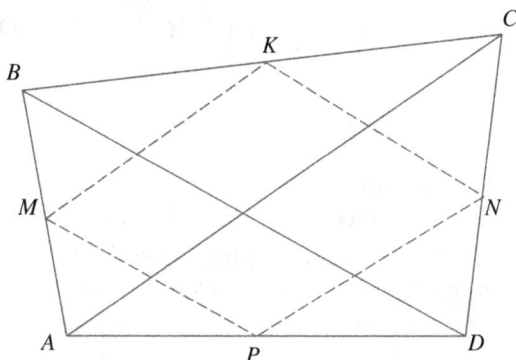

We are required to prove a statement that connects the squares of the certain segments.

Should this lead us, for instance, in the direction of utilizing the vector algebra techniques? If we work with sums of vectors expressing the vectors-diagonals through the vectors-sides and introduce a vector scalar product (dot product), this may lead us to a desired result. Is this a good idea? Can we make a reasonable prognosis about where this idea should take us?

This is perhaps one of the possible directions for contemplating a plan for a solution.

However, if we take this approach, we can't clearly see the path to the ultimate goal; and besides, it might involve tedious calculations. So, this does not look like a promising suggestion.

On the other hand, by analyzing the picture, we should notice that $MK$ is the midline in the triangle $ABC$ and $PN$ is the midline in the triangle $ADC$. Recalling the property of the midline of a triangle, this implies that each segment $MK$ and $PN$ is parallel to $AC$ and is half as long as $AC$. Similarly, we get that in the triangles $ABD$ and $CBD$, $MP \parallel BD \parallel KN$ and $MP = KN = \frac{1}{2}BD$. Thus, it becomes obvious that $MKNP$ is a parallelogram, and we should be able to make a reasonable prognosis to apply the properties of a parallelogram in the undertaking of our further steps. It's a great idea. Since the pairs of the opposite sides of $MKNP$ are parallel and equal, it is indeed a parallelogram. It looks like we are now moving in the right direction.

We see that

$$MK^2 + KN^2 + NP^2 + PM^2$$

$$= 2MK^2 + 2KN^2 = 2\left(\left(\frac{1}{2}AC\right)^2 + \left(\frac{1}{2}BD\right)^2\right)$$

$$= \frac{1}{2}AC^2 + \frac{1}{2}BD^2,$$

which is the desired result.

In the solution, once you have $MK = PN = \frac{1}{2}AC$ and $MP = KN = \frac{1}{2}BD$ you can just plug these in to verify the equation. The fact that $MKNP$ is a parallelogram doesn't really matter. But, this is a very interesting bypassing result; we proved that in a convex quadrilateral the midpoints of its sides are the vertices of a parallelogram.

Obviously, the property will hold for a concave quadrilateral as well. It can be easily proved in a similar fashion as for a convex quadrilateral, and we leave this exercise for the readers to justify.

Furthermore, extending the study of problems-relatives, we can draw an analogy with the considered property for a coplanar quadrilateral and a skew quadrilateral, a quadrilateral not contained in a plane. That is,

> *A quadrilateral formed by joining the middle points of the adjacent sides of a skew quadrilateral is a coplanar parallelogram.*

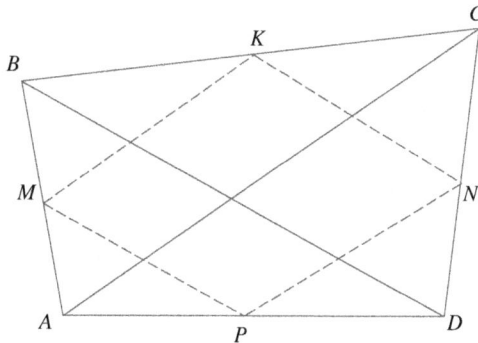

Assume now that $ABCD$ is a skew quadrilateral. Once again, we will apply the properties of a midline of a triangle and make the same conclusions that in the triangles $ABD$ and $CBD$, $MP \parallel BD \parallel KN$ and in the triangles $ABC$ and $ADC$, $MK \parallel AC \parallel PN$. Since parallel lines lie in the same plane, we may conclude that there is a plane passing through $MP$ and $KN$ and there is a plane passing through $MK$ and $PN$. Both, the planes through $MP$ and $KN$ and through $MK$ and $PN$, pass through the same four points $M$, $K$, $N$, and $P$. Therefore, it is evident that the two planes must coincide because a plane is uniquely defined by any three non-collinear points. Thus, the quadrilateral $MKNP$ is coplanar. Again, as proved, its opposite sides are parallel and have the same length. Therefore, the quadrilateral $MKNP$ is a parallelogram. So, we see that the considered property is valid for any type of a quadrilateral. As a consequence, it follows that the midpoints of the sides of a parallelogram (it is a quadrilateral) will be the vertices of a parallelogram as well. We can pose an intriguing general question about the type of quadrilateral that is obtained by connecting the points that divide the sides of a parallelogram not in half but in any ratio (same ratio for all sides, going, for example, clockwise); will it be a parallelogram as well? In fact, another achievement in solving the original problem was to "invent" the following new problem.

**Problem 16.** Prove that the points that divide the sides of a parallelogram in the same ratio going in the same direction, clockwise, for example, are the vertices of a parallelogram.

**Solution.** Assume that the points $M$, $K$, $N$, and $P$ are selected on the sides of the given parallelogram $AB$, $BC$, $CD$, and $DA$, respectively, such that $\frac{MB}{AB} = \frac{KC}{BC} = \frac{ND}{DC} = \frac{AP}{AD}$.

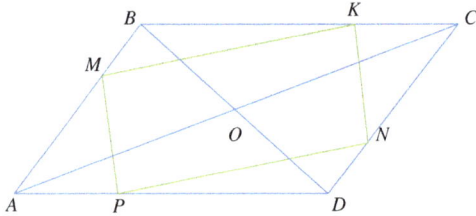

We have to prove that $MKNP$ is a parallelogram. This goal will be easily achieved by applying the central symmetry with the center $O$, the point of intersection of the diagonals of $ABCD$. Since $O$ divides the diagonals of $ABCD$ in half, then in our central symmetry the image of $A$ is $C$, the image of $B$ is $D$, and we see that segment $AB$ will be taken into $CD$. It implies that every point on $AB$ will be transferred into a respective point on $CD$, that is, the image of $M$ will be the point $N$ on $CD$, such that it divides $CD$ in the same ratio as $M$ divides $AB$. In the same fashion, we obtain that $K$ will be transferred into $P$. By the properties of the central symmetry, $MP = KN$ and $MP \parallel KN$, $MK = PN$ and $MK \parallel PN$. Therefore, $MKNP$ is a parallelogram, as it was required to be proved.

It merits now to look back and review how we got to this interesting result.

We started with the problem about a convex quadrilateral that has a specific property connecting its diagonals with its midlines.

By analyzing the given information and making a proper prognosis about how to approach this problem, we came across an auxiliary parallelogram. This helped us to establish the required relationship. But, it also revealed another interesting property that is valid for any quadrilateral; that is, by connecting the midpoints of its sides, we form a parallelogram. Going from a general case for any quadrilateral, we concluded that this property is valid for parallelograms as well. Even though this observation had nothing to do with the original problem we were solving, it allowed us to make an interesting generalization of a new problem. We posed a question about a problem-relative and considered a general case scenario for getting a new parallelogram by connecting the points that divide the sides of the original parallelogram not just in half but in any $\frac{m}{n}$ ratio, the same ratio for all sides.

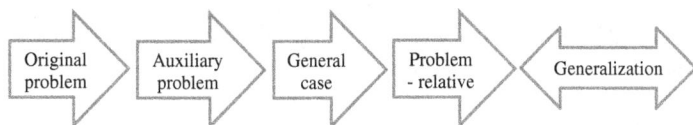

To summarize, through making a proper prognosis for justifying our steps in a solution process, evaluating an auxiliary problem and intermediate results, going from a general case to a specific scenario (from any quadrilateral to a parallelogram), and looking for a relative of a new problem, we extended a specific case to arrive at its interesting generalization.

Generalization means looking for a bigger picture and broader patterns. By using generalizations, we extend the horizon and may indicate the broader range of objects satisfying a specific property.

A good example of investigating a more general case would be to go back and analyze the outcome from our Problem 4 discussed at the beginning of the chapter. We may ponder the following question: what if the hypotenuse had been $c$ times the altitude for some number $c \neq 4$? Then, our method would not work. It was fine for the very specific numerical value when the hypotenuse is four times the length of the altitude dropped to it. Can we extend this problem and solve a more general problem?

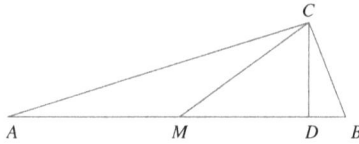

To simplify calculations, let's choose the altitude $CD$ to be our unit of distance; then, $CD = 1$ and, respectively, $AB = CD \cdot c = c$. Denote $AD = x$ and $DB = y$. Since $\triangle ADC$ and $\triangle CDB$ are similar, then $\frac{CD}{AD} = \frac{DB}{CD}$, or in our nominations, $\frac{1}{x} = \frac{y}{1}$, and therefore $xy = 1$. Also, we have that $AD + DB = x + y = AB = c$. Combining these equations, we get the system

$$\begin{cases} xy = 1, \\ x + y = c, \end{cases}$$

from which we arrive at the quadratic equation $x^2 - cx + 1 = 0$. Solving this equation for $x$ gives $x_1 = \frac{c + \sqrt{c^2 - 4}}{2}$, $x_2 = \frac{c - \sqrt{c^2 - 4}}{2}$. Observing that clearly $c > 0$, it follows that $c \geq 2$ (this results from finding the permissible domain of values of $c$ satisfying $\sqrt{c^2 - 4}$).

Our goal was to find the angle values in $\triangle ACB$. Having $AD = x$, we can get the value of $\angle BAC = \angle DAC$ from the right triangle $ADC$ expressing its tangent through $x$ as $\tan \angle DAC = \frac{CD}{AD} = \frac{1}{x}$. Substituting the values of $x$ into the last expression, we get two different solutions for $\angle DAC$ and, respectively, for $\angle BAC$ as $\angle BAC = \arctan\left(\frac{1}{x}\right)$. The second angle $\angle ABC$ now can be easily calculated as the difference of $90°$ and $\angle BAC$.

Recalling the material discussed at the beginning of this chapter, it merits noting that Thales's Theorem is a special case of the following more general statement known as the Central Angle Theorem:

Given three points $A$, $B$, and $C$ on a circle with center $O$, the angle $\angle AOC$ is twice as large as the angle $\angle ABC$, or in other words, *the measure of inscribed angle $\angle ABC$ is always half the measure of the central angle $\angle AOC$.*

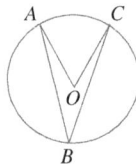

This theorem is important in geometry, but easy to prove. The readers are invited to prove it independently.

The art of making a generalization of an obtained result is perhaps one of the most valuable in mathematics. It is always very compelling and delighting to find a positive response to an inquiry that can lead to a more general statement than the examined one. It becomes even more impressive when you come across a generalization of a well-known problem or theorem.

We were referring to the Pythagorean Theorem in this chapter during our discussion of the converse problems. We will be using this famous theorem one more time now by illustrating an interesting generalization of the theorem's statement. What is especially appealing in our following observations is the fact that we will derive the generalization after restating the statement of the Pythagorean Theorem (recall our discussion about usefulness of this technique!). The theorem asserts that

> *In a right triangle the square of the hypotenuse is equal to the sum of the squares of the two legs: $c^2 = a^2 + b^2$.*

Since ancient Greek times, this theorem had a different statement and it was formulated as

> *The area of the square built on the hypotenuse is equal to the sum of the areas of the squares built on the two legs.*

We will be using the last statement and will analyze it here.

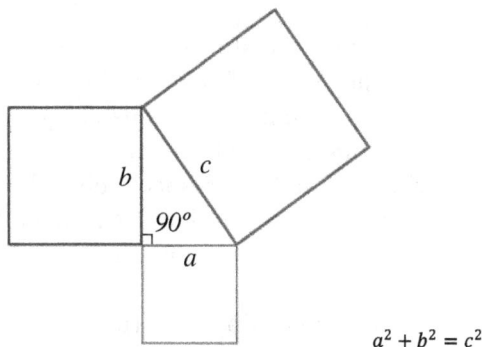

$$a^2 + b^2 = c^2$$

Obviously, any squares are similar to each other. So, is it just a coincidence that the area of the big square built on the hypotenuse is equal to the sum of the areas of two other similar squares, or this property would work for any similar polygons, and we can get a generalization of the famous theorem? In other words, is this a correct statement:

*If we construct similar polygons with corresponding sides on the sides of a right triangle, then the sum of the areas of the ones on the two legs equals the area of the one on the hypotenuse?*

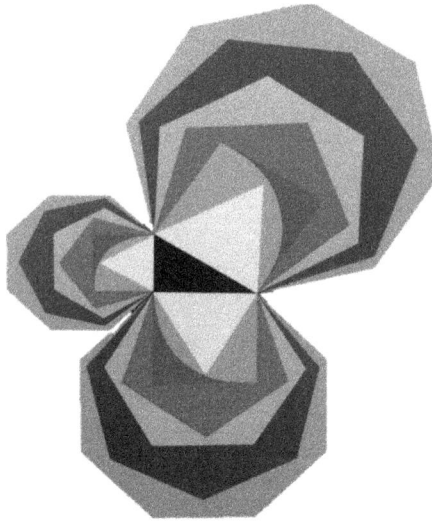

The elegant way to prove the above statement is by applying the Theorem of Ratios of the Areas of Similar Polygons. It states that the ratio of the areas of two similar polygons is equal to the squared ratio of their corresponding linear elements.

Let's denote by $a$, $b$, and $c$, respectively, the legs and the hypotenuse of the right triangle, and by $S_1, S_2$, and $S_3$ the areas of the respective similar polygons built on the sides of the right triangle.

According to the Theorem of Ratios of the Areas of Similar Polygons, we get that

$$\frac{S_1}{a^2} = \frac{S_2}{b^2} = \frac{S_3}{c^2}.$$

It follows that $S_1 = S_3 \cdot \frac{a^2}{c^2}$ and $S_2 = S_3 \cdot \frac{b^2}{c^2}$. Adding the last two equalities and recalling that $c^2 = a^2 + b^2$ gives

$$S_1 + S_2 = S_3 \cdot \frac{a^2}{c^2} + S_3 \cdot \frac{b^2}{c^2}$$

$$= S_3 \cdot \left( \frac{a^2}{c^2} + \frac{b^2}{c^2} \right)$$

$$= S_3 \cdot \frac{a^2 + b^2}{c^2}$$

$$= S_3 \cdot \frac{c^2}{c^2} = S_3.$$

Hence, we indeed obtained that the sum of the areas of the similar polygons built on the two legs is equal to the area of the similar polygon built on the hypotenuse, proving that the suggested generalization of the Pythagorean Theorem is valid.

Looking for siblings-problems as relatives of the same family of problems, restating a problem's conditions, solving converse problems, applying inversion technique, and experimenting with the given original information by modifying it and posing questions regarding possible generalizations help in better understanding of the considered problem, enhance problem-solving skills, and greatly stimulate ingenuity and creativeness which are so important in developing of mathematical thinking.

# Chapter 10

# Alternative Solutions Search

One of the great exercises helping to improve your skills in problem-solving is the search for alternative solutions to a problem. By making different approaches to a problem, one develops critical thinking, sharpens problem-solving techniques, and gets a deeper understanding of the issues involved and possible links and connections among various math topics. To become a good mathematician, one should not be satisfied with one solution only, but rather should be trying to find alternative shortest, most elegant, and efficient solutions. As we find a new and better solution, we may come across new and interesting facts. Comparing the completed solutions, reevaluating the results and pathfinding, we not merely consolidate our knowledge, but stimulate also important discoveries and observations not related to the studied problem at all at first glance.

In this chapter, we will examine several problems and demonstrate different points of view for their solutions; we will compare and analyze them.

**Problem 1.** A mother is 35 years old and her son is 5 years old. In how many years will the mother be twice as old as her son?

**Solution.**

**1. Arithmetic solution.** The year in which the mother becomes twice as old as the son, the difference of their ages should equal the son's age. Since the difference in ages is a constant number and equals $35 - 5 = 30$, we can conclude that the son will be 30 years old in that year. His mother will be 60 years old, and it will happen in $60 - 35 = 25$ years (or you find the same number as $30 - 5 = 25$ using the son's age now and in the year in question).

**2. Algebraic method.** The above reasoning has vivid interpretation through utilizing variables to set up the equation satisfying the problem's conditions. Assume the mother becomes twice as old as her son in $x$ years. Her age then will be $(35 + x)$, the age of her son will be $(5 + x)$, and according to the problem's statement,

$$2 \cdot (5 + x) = 35 + x.$$

Solving this simple equation, we get $x = 25$.

**3. Intelligent guess method.** At this point, I imagine the ambitious readers would raise eyebrows in surprise. A mathematician should never guess! Yes, it is true that we are always looking for rigorous proof. However, sometimes a good guess, even a wrong one, may clarify the picture and eventually lead to a good result. By proving or rejecting your assumption, you may come across important observations or facts paving the way to the problem's solution. By "intelligent guess", we mean some useful analysis, assumption, or even a guess that would lead to a valid solution. Another argument in support of the guess-and-choose technique is obtaining a quicker result under specific circumstances. On many tests, multiple-choice questions can be solved much faster by applying this particular technique. Guess the correct answer and verify if it works.

Keep in mind that we are looking for a natural number as a solution to the problem. It significantly delimits the area search. Also, the age at which mother becomes twice as old as the son has to be an even number. It should be natural to pick as the first choice the year when she will be 50 years old. It will happen in 15 years, so the son will be 20 years old. We see that it is not a correct guess because $50 \neq 20 \cdot 2$. The second selection of the mother's desired age at 60 gives the correct result. Indeed, in the year she becomes 60 years old, which will happen in $60 - 35 = 25$ years, the son will be $5 + 25 = 30$ years old. Since $60 = 2 \cdot 30$, we are done.

**Answer:** In 25 years, the mother will be twice as old as her son.

Mixture problems involve creating a mixture from two or more things, and then determining some quantity (percentage, price, etc.)

of the resulting mixture. They are usually solved by interpreting a problem's data into algebra terms and setting one or several equations. Solving an equation or a system of equations gives the desired results. However, there is another interesting technique for simplifying solutions of this type of problems. Let's consider both alternatives here.

**Problem 2.** How many liters of a 30% acid solution have to be mixed with a 10% acid solution to get 10 liters of a 15% solution?

**Solution.**

**1. Algebraic method.** Let $x$ be the number of liters of a 30% solution and let $y$ be the number of liters of a 10% solution needed to get 10 liters of the 15% acid solution.

So, the first equation is $x + y = 10$ liters. To get the second equation, we have to interpret the acid solution concentration in the 15% desired solution through the mixture of $x$ liters of a 30% acid with $y$ liters of a 10% acid. In $x$ liters, 30% acid has weight of $0.3x$ liters, and in $y$ liters, 10% acid has weight $0.1y$ liters. They have to be mixed to get the new 10-liter concentration in which 15% acid would have $10 \cdot 0.15 = 1.5$ liters. The liters of acid from the 30% solution plus the liters of acid in the 10% solution add up to the liters of acid in the 15% solution, which can be expressed as $0.3x + 0.1y = 1.5$. Our analysis is conveniently represented by the table below.

|  | Liters in solution | Percent acid | Total liters acid |
|---|---|---|---|
| 30% Solution | $x$ | 30% | $0.3x$ |
| 10% Solution | $y$ | 10% | $0.1y$ |
| Mixture | $x + y = 10$ | 15% | $0.15 \cdot 10 = 1.5$ |

This leads to the following system of two linear equations:

$$\begin{cases} x + y = 10, \\ 0.3x + 0.1y = 1.5. \end{cases}$$

From the first equation, $x = 10 - y$. Using this, we can substitute for $x$ in the second equation and solve it for $y$.

$$0.3(10 - y) + 0.1y = 1.5,$$
$$3 - 0.3y + 0.1y = 1.5,$$
$$0.2y = 1.5,$$
$$y = 7.5.$$

We need 7.5 liters of the 10% solution, and finding $x$ from the first equation, we get $x = 10 - y = 10 - 7.5 = 2.5$ liters of the 30% solution.

**2. Arithmetic solution.** Let's consider the following scheme: write one under the other the given acid concentration percentages, 30% and 10%, then put to the left the required 15% in the new mixture. Find the difference between the greater percentage concentration of 30% and the required percentage of 15% and write it to the right of the smaller percentage of 10%; find the difference between the required mixture percentage of 15% and the smaller percentage concentration of 10% and write it to the right of the greater percentage of 30%.

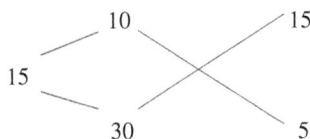

We arrive at the conclusion that $15 : 5 = 3$ times more liters of a 10% acid have to be added than a 30% acid to get 10 liters of the new 15% acid, which is 7.5 liters of the 10% acid and 2.5 liters of the 30% acid. This is the same result as we got using the algebraic techniques.

I find the second alternative solution to be very attractive in solving any similar mixture problems. It is not hard to justify the general case using the following analysis.

Assume we are mixing two quantities (items, percentages, cost, etc.) $a$ and $b$. To be more specific, let's say, we are mixing two different products with the price \$$a$ per item and \$$b$ per item and $a < b$. The goal is to get the mixture of both products at price \$$c$ per item.

Obviously, if $c < a$ or $c > b$, the problem has no solutions (mixing two cheap products, one can't get an expensive new product; likewise, mixing two expensive products, one can't get a cheap new product). Without loss of generality, we can say that $a < c < b$. Let's mix 1 item of the first product with $x$ items of the second product. We will get $(1 + x)$ items with the price of $(a + bx)$ dollars. We want the new mix to be sold at $c$ dollars per item. Therefore, the quantity $x$ of the second product to be used in a mixture is to be determined from the equation

$$a + bx = c \cdot (1 + x).$$

Solving the last equation for $x$ gives $x = \frac{c-a}{b-c}$. So, the products have to be mixed in a ratio $1 : \frac{c-a}{b-c}$ or equivalently, $(b - c) : (c - a)$. But, this is exactly the outcome represented by the scheme we applied in the alternative arithmetic solution above:

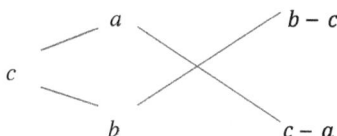

**Problem 3.** Solve the equation $(2x - 1)\sqrt{x - 2} = 7\sqrt{2}$.

The straightforward approach is to get rid of irrationality on the left-hand side of the equation by squaring both sides, and then continue solving the cubic equation. We suggest readers follow this path and find the solutions. Is it the best method for solving this equation? Perhaps it is not.

*Intelligent guess* here may be very useful in much simplifying the solution. First, determining the domain of this equation, we see that $x > 2$ since the expression under the square cannot be negative and $x \neq 2$ because otherwise the left-hand side becomes equal to 0 and $0 \neq 7\sqrt{2}$. The way this equation is written may lead to an assumption that the factors with radical signs on both sides should be equal. If that's the case, then we need to consider the following system of equations:

$$\begin{cases} 2x - 1 = 7, \\ \sqrt{x - 2} = \sqrt{2}. \end{cases}$$

Solving each equation, we obtain the common solution $x = 4$. Let's prove now that the original equation has no other solutions.

Indeed, if $x > 4$, then obviously, $(2x - 1)\sqrt{x - 2} > 7\sqrt{2}$.

If $2 < x < 4$, then $(2x - 1)\sqrt{x - 2} < 7\sqrt{2}$. So, in either case, we see that there are no other solutions than 4.

I have no doubt that after completing the straightforward solution, the readers should much appreciate having the alternative in the intelligent guess technique, which gives a much more elegant and easy solution in this case.

Never underestimate a good guess and helpful idea. If you are lucky and the idea looks advantageous, go for it; just make sure you eventually validate all you steps and get the rigorous proof that your assumption was correct.

**Problem 4.** Solve the equation $|x - 7| = 3$.

**Solution.**

**1. Algebraic solution.** We know that

$$|x| = \begin{cases} x, & \text{if } x \geq 0, \\ -x, & \text{if } x < 0. \end{cases}$$

So, to solve the given equation, we have to consider two scenarios.

1. When $x \geq 7$, $|x - 7| = x - 7$, and we get the equation $x - 7 = 3$, from which $x = 10$. This is our first root of the equation.
2. When $x < 7$, $|x - 7| = -x + 7$, and the equation simplifies to $-x + 7 = 3$, solving which we get $x = 4$. This is the second root of the equation.

**Answer:** 4, 10.

**2. Geometrical solution.** An absolute value is the distance from zero to the number on the number line, or in other words, the positive version of the number.

Let's state our problem in words: find the value of variable $x$ such that the absolute value of the difference of $x$ and 7 equals 3. It can be interpreted as finding all such numbers that are equidistant on a number line from 7 by 3 units. So, let's consider a number line and find all the points located from 7 at the distance of 3 units.

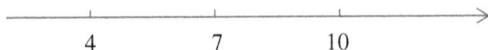

Obviously, there are two such points, 4 and 10, that are equidistant from 7 by 3. The equation has two solutions, $x = 4$, $x = 10$.

**Problem 5.** Solve the equation $|x - 1| + |x + 2| = 3$.

**Solution.**

**1. Algebraic solution.** Using the definition of the absolute value of a number, we will now need to consider three scenarios.

1. When $x \geq 1$, we have $x - 1 + x + 2 = 3$. Solving this equation gives $x = 1$.
2. When $-2 \leq x < 1$, we have $-x + 1 + x + 2 = 3$. This becomes $0 \cdot x = 0$ and clearly, every number from the interval $[-2, 1[$ satisfies the equation.
3. When $x < -2$, we get $-x + 1 - x - 2 = 3$, from which $x = -2$. This should be rejected because we consider solutions only for $x < -2$.

Combining the results from the above analysis, we see that any number from the interval $[-2, 1]$ is the solution of the given equation.

**2. Geometrical solution.** Similar to Problem 4, we can state that solving our equation reduces to finding all numbers such that the sum of the distances from them to $-2$ and $1$ equals 3.

Once again, let's consider the number line and locate points $-2$ and $1$ on it.

Obviously, selecting any point between $-2$ and $1$, including the end points, we see that the sum of the distances from the selected point to end points of the segment equals 3. This is true because the length of this segment equals 3. Therefore, the solutions of the given equation are all numbers from the segment $[-2, 1]$, the same result as we got in our algebraic solution.

Considering various alternative solutions to a problem proves useful in applying the obtained results and conclusions in solving similar related problems. For example, bearing in mind the results obtained in problems 4 and 5, we can apply them to tackle the following Olympiad-type problem:

**Problem 6.** Given the equation $|| \cdots |x - a_1| - a_2| - \cdots | - a_{10}| = a$, where $a$ and $a_i (i = 1, 2, \ldots, 10)$ are some real numbers, determine what is the maximum possible number of its roots.

**Solution.** Solving any simple equation of the type $|x - a| = b$, we have the maximum of two roots depending on values of $a$ and $b$ (if neither of them equals 0). So, after "opening" the first absolute value sign, there are a maximum of two solutions, or $2^1$. After "opening" the second absolute value sign, there will be a maximum of four solutions, two for each of the two obtained equations, or $2^2$. Applying a similar logic in every next step, we arrive at the conclusion that there are a maximum of $2^{10}$ possible solutions of the original equation.

**Problem 7.** In the given triangle $ABC$ with the sides $BC = a$ and $AC = b$, the medians dropped to these sides are perpendicular. Find the length of the third side $AB$.

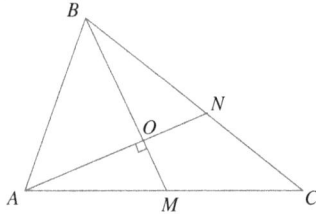

**Solution 1.** Draw the medians $BM$ and $AN$, and denote by $O$ the point of their intersection. It is given that $BM \perp AN$. Hence, we got three right triangles $AOB$, $AOM$, and $BON$ with the right angles formed by the vertex $O$. Our goal is to find $AB$. We apply the Pythagorean Theorem to triangle $AOB$ to obtain its hypotenuse $AB$, $AB^2 = AO^2 + BO^2$. So, if we manage to find $AO$ and $BO$, or the sum of their squares (it's always important to keep in mind the final question asked and get directly to it, if possible), the problem will be solved. Let's analyze the given conditions and see how we can apply them in the direction for the conceived plan.

BM and $AN$ are the medians of the triangle, which trisect each other at their point of intersection, the triangle's centroid $O$. So, $AO = 2ON$ and $BO = 2OM$. Consider now two right triangles $AOM$ and $BON$ and express from them the squares of $AO$ and $BO$ in terms of the given sides $BC = a$ and $AC = b$.

In the right triangle $AOM$, $AM^2 = AO^2 + OM^2 = AO^2 + \frac{1}{4}BO^2$, or substituting $AM = \frac{1}{2}AC = \frac{b}{2}$, we can rewrite the equality as $\frac{b^2}{4} = AO^2 + \frac{1}{4}BO^2$. After multiplying both sides by 4, it simplifies to

$$4AO^2 + BO^2 = b^2. \tag{1}$$

In the right triangle $BON$, we get $BN^2 = BO^2 + ON^2 = BO^2 + \frac{1}{4}AO^2$.

Substituting $BN = \frac{1}{2}BC = \frac{a}{2}$, we can rewrite the equality as $\frac{a^2}{4} = BO^2 + \frac{1}{4}AO^2$ and after multiplying both sides by 4, it simplifies to

$$4BO^2 + AO^2 = a^2. \tag{2}$$

Adding (1) and (2) gives $5BO^2 + 5AO^2 = a^2 + b^2$, from which

$$BO^2 + AO^2 = \frac{a^2 + b^2}{5}. \tag{3}$$

Recall that in the right triangle $AOB$ its hypotenuse $AB$ can be found from the equality $AB^2 = AO^2 + BO^2$. Substituting the sum of the squares of $AO$ and $BO$ from (3) gives $AB^2 = \frac{a^2 + b^2}{5}$, and we finally arrive at $AB = \sqrt{\frac{a^2 + b^2}{5}}$.

It's worth emphasizing that, in this solution, we skipped calculating the length of $BO$ and $AO$, and conveniently obtained the outcome for $AB$ from manipulating the system of two equations without even solving it.

**Solution 2.** Let's draw the third median $CK$ to the side $AB$. It must pass through the centroid $O$. Denote the length of the unknown side $AB = x$. In the right triangle $AOB$, $OK$ is the median dropped from the right angle's vertex to the hypotenuse; therefore, $AK = KB = KO = \frac{x}{2}$ (see the previous chapter).

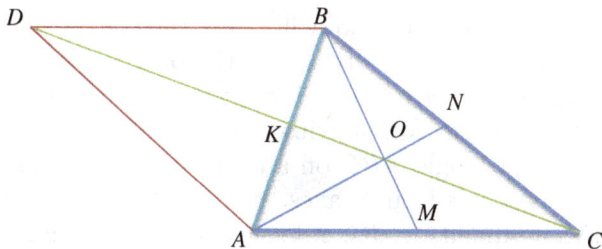

For the second solution, we will extend $CK$ and find $D$ located on its extension, such that $DK = KC$ (in fact, we find the image of $C$ in the central symmetry with the center $K$). Connecting $D$ with $A$ and $B$, we obtain a supplemental parallelogram $DBCA$ (it is a parallelogram because its diagonals $CD$ and $AB$ by construction bisect each other at their point of intersection $K$), the properties of which will help us to find $AB$.

Why did we do the additional construction? How can it help us in finding $AB$? Since the opposite sides of a parallelogram are equal, then $DB = AC = b$ and $DA = BC = a$, and we can express $DC$ in terms of $x$, $DC = 2KC = 2 \cdot 3KO = 6 \cdot \frac{x}{2} = 3x$ (since the centroid $O$ splits the median $KC$ in the triangle $ABC$ in a ratio 2:1).

Now, by utilizing the property of a parallelogram that the sum of the squares of its diagonals equals the sum of the squares of its sides, we can state that

$$2BC^2 + 2AC^2 = AB^2 + DC^2.$$

Substituting the respective values, we obtain that $2a^2 + 2b^2 = x^2 + (3x)^2$, which simplifies to $10x^2 = 2a^2 + 2b^2$ and finally,

$$x = \sqrt{\frac{a^2 + b^2}{5}}.$$

We succeeded in solving our problem thanks to two remarks. First, we invented an advantageous auxiliary problem introducing a new figure, a parallelogram. Second, we discovered a path from the auxiliary to the original problem by using a compelling property of a parallelogram connecting the lengths of the sides with the lengths of its diagonals. The trick with auxiliary parallelogram is worth remembering because it proves useful in many similar problems involving calculations of sides or medians in a triangle.

**Solution 3.** In this solution, we will apply the Cartesian coordinate plane technique. Since there is no restriction on how to place our triangle on the coordinate plane, then it is up to us to do it in the most advantageous way for carrying out our plan.

We will place triangle $ABC$ on a coordinate plane is such a way that point $A$ is located on $Y$-axes, point $B$ is located on $X$-axes, point $N$, the midpoint of $BC$, is located on $Y$-axes, and point $M$, the midpoint of $AC$, is located on $X$-axes. So, we have $AN \perp BM$, as it is given in the problem, $AN$ and $BM$ are the medians in $\triangle ABC$, and $BC = a$ and $AC = b$.

Now, we will assign the coordinates of the points $A$ and $B$ as $A(0, -y)$, $B(x, 0)$. Then, the lengths of the segments $BO$ and $AO$ can be expressed as $BO = x$, $AO = |-y| = y$.

Draw $CL \perp OX (L \in OX)$ and $CF \perp OY (F \in OY)$. The right triangles $AOM$ and $CLM$ are congruent because $\angle LCM = \angle OAM$ as interior angles by the parallel lines $CL$ and $OA$ and the transversal $AC$, $\angle LMC = \angle OMA$ as vertical angles, and $CM = MA$ ($M$ is the midpoint of $AC$).

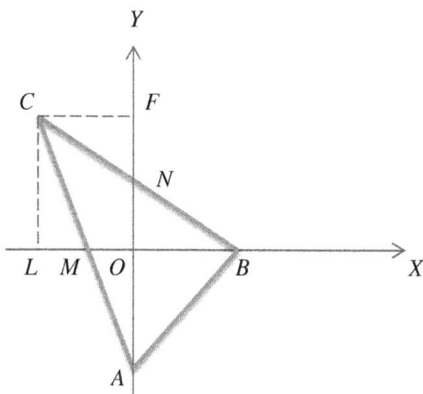

Therefore, $CL = OA$. In a similar fashion, we can prove that from the congruency of the right triangles $CFN$ and $BON$, it follows that $CF = BO$. It implies that the coordinates of $C$ are $-x$ and $y$, $C(-x, y)$.

Having the coordinates of the points $A(x_A, y_A)$ and $B(x_B, y_B)$ in a two-dimensional space, the distance between them is calculated by the general formula as $AB = \sqrt{(x_B - x_A)^2 + (y_B - y_A)^2}$. So, in our nominations, we have $AB = \sqrt{x^2 + y^2}$.

Expressing the lengths of the given sides $BC = a$ and $AC = b$ through the coordinates, we have

$$BC^2 = (-x - x)^2 + (y - 0)^2 = 4x^2 + y^2, \quad \text{or} \quad a^2 = 4x^2 + y^2 \quad (1)$$

and

$$AC^2 = (-x - 0)^2 + (y - (-y))^2 = x^2 + 4y^2, \quad \text{or} \quad b^2 = x^2 + 4y^2. \quad (2)$$

Adding (1) and (2) gives $a^2 + b^2 = 5x^2 + 5y^2$, from which we get $x^2 + y^2 = \frac{a^2+b^2}{5}$. Therefore, $AB = \sqrt{x^2 + y^2} = \sqrt{\frac{a^2+b^2}{5}}$.

The Cartesian coordinates method is a powerful tool allowing the connection of the coordinates of the points with calculation of segments' lengths. Our decision on how to position the given triangle in the coordinate plane was the first and a conspicuous step. It allowed utilizing the given information in the most advantageous way. The subsequent additional constructions provided the link between the given triangle and auxiliary triangles allowing the calculation of the distances between points by their coordinates in a relatively simple and elegant way.

While solving a problem, it is important to always keep in mind the question asked and, if possible, get to the answer by skipping the intermediate steps. This was accomplished in the considered Solution 3. Similar to Solution 1, we arrived at the desired result directly from the system of two equations without solving it for $x$ and $y$.

**Problem 8.** Given a circle with an invisible center, find the location of the circle's center using a compass and a straightedge.

**Solution 1.**

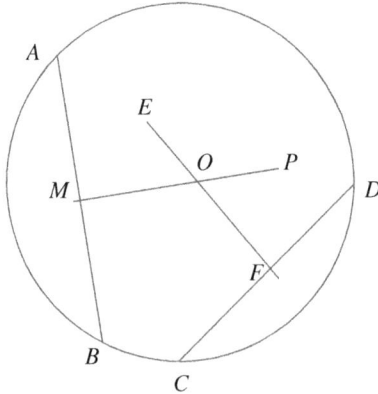

Dealing with a circle, it's natural to bring into play a circle's properties, many of which emanate from a circle's definition as the locus of points on a plane equidistant from the given point, its center. Let's pick four random points $A$, $B$, $C$, and $D$ on the circle, such that connecting $A$ and $B$, and $C$ and $D$, we get two non-parallel chords $AB$ and $CD$. Being located on the given circle, $A$, $B$, $C$, and $D$ are equidistant from the invisible circle's center in question. The set of all points that are at the given distance from $A$ and $B$ is the perpendicular bisector $MP$ to the segment $AB$. Likewise, the set of all points that are at the given distance from $C$ and $D$ is the perpendicular bisector $FE$ to the segment $CD$. Since we constructed two non-parallel chords, their perpendicular bisectors will necessarily intersect. The point of their intersection $O$ is equidistant from all four points and, therefore, is the sought-after center of the given circle.

**Solution 2.**

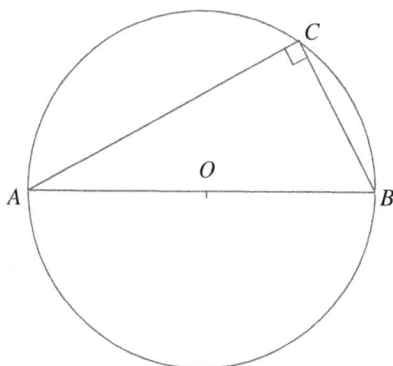

In this solution, we will exploit the property that all central angles in a circle subtended by the diameter are right angles. First, we draw an arbitrary chord $AC$. The next step is to construct $CB \perp AC$ at point $C$ till its intersection with the circle at $B$. By connecting $A$ and $B$, we get the diameter of the given circle $AB$. Indeed, by construction, angle $ACB$ is the right angle. Therefore, the hypotenuse $AB$ of the right triangle $ACB$ is the diameter of the circumcircle of the right triangle $ACB$. The final step is to locate the midpoint $O$ of $AB$, which is the desired center of the given circle.

It's noteworthy that $O$ can be located not only as the intersection of the perpendicular bisector to $AB$ with $AB$ but alternatively as the intersection of the diagonals of the rectangle formed by constructing parallel lines to $BC$ at $A$ and to $AC$ at $B$ till their intersections with the circle. We leave these simple constructions for the readers to complete. As we mentioned before, we hope that readers will carefully and fully perform all the standard constructions, justifications, and rigorous proofs of the intermediate steps.

**Solution 3.** Contrary to the first step in our Solution 1, now we draw two random parallel chords, $AB \parallel CD$. We consider the selection of $A$, $B$, $C$, and $D$ such that $ACDB$ is a trapezoid, not a rectangle, which was, basically, already considered in our Solution 2 in the final step in the alternative "extension" of the right triangle into a rectangle, the diagonals of which intersect at the desired center of the given circle. Denote by $N$ the point of intersection of the diagonals $AD$ and $BC$ and extend the non-parallel sides $AC$ and $BD$ till their intersection at $P$. Being inscribed in the circle, $ACDB$ is an isosceles trapezoid with $AC = BD$.

As the result of our constructions, the formed triangles $APB$ and $CPD$ are isosceles with the respective equal sides $AP = BP$ and $CP = DP$.

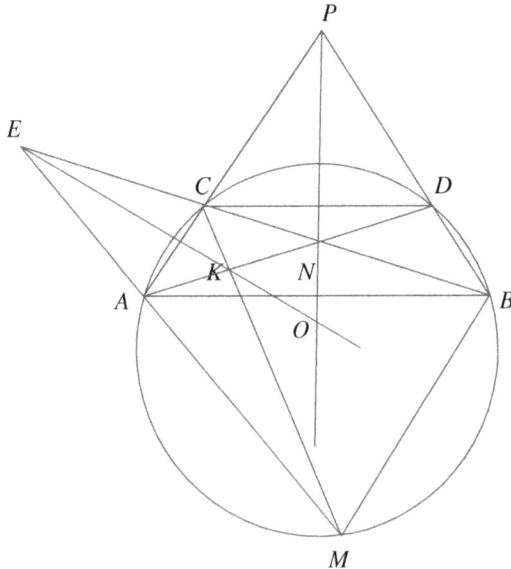

The straight line passing through $P$ and $N$ contains the median of each isosceles triangle (recall Jacob Steiner's problem we have used quite a few times during the course of the book about the properties of the line passing through the points of intersection of diagonals and extended non-parallel sides of a trapezoid). There-fore, it contains the altitude of each triangle as well, and we can conclude that $PN$ is the locus of all points equidistant from $A$, $C$, $D$, and $B$. Now, our goal is to locate the center of the given cir-cle on $PN$. It can be easily achieved by making similar observa-tions and doing construction of another trapezoid inscribed in the given circle. For example, let's draw $BM \parallel CA$ till its intersection at $M$ with the given circle. $BMAC$ is an isosceles trapezoid with $MA = BC$. Draw its diagonals $BA$ and $MC$ and denote their point of intersection by $K$. Next, extend the non-parallel sides $MA$ and $BC$ till their intersection at $E$. Finally, draw $EK$ and find $O$, its point of intersection with $PN$. $O$ is the desired center of the given circle.

Someone might find the last solution more complicated and even cumbersome in comparison with the other two solutions. However, it suggests a different approach to the problem's solution, which heavily relies on properties of a trapezoid and isosceles triangle. Furthermore, it may ignite the exploration of various modifications of the problem's data and devising your own new problems–challenges. The previous two chapters were devoted to studying this very important and use-ful problem-solving developing technique — modifying a problem's attributes and generating new problems based on the already solved ones. The considered problem can serve this purpose as well. By applying the obtained results, we can formulate, for example, the following problem:

> *There is given a circle with an invisible center. Points A, C, D, B, and M on this circle are such that AB $\parallel$ CD and BM $\parallel$ CA. Find the location of the center of the circle using a straightedge only (no compass use is allowed).*

Most likely, this should be considered as a non-conventional and tricky challenge. But, in fact, the solution of it is exactly the same as we just demonstrated above. There is no need to use a compass to get our constructions done. The main idea behind the solution will be in identifying and constructing two isosceles trapezoids $ACDB$ and $BMAC$. All the next steps will be drawing straight lines using a straightedge only while connecting the vertices of trapezoids obtain-ing the points of intersection of their diagonals, extending the non-parallel sides, and getting their points of intersections, and finally,

locating the invisible center as the point of intersection of straight
lines passing through the points of intersection of diagonals and
extended non-parallel sides for each trapezoid.

It will be a good exercise to come up with your own new problems
contemplated from that considered above.

**Problem 9.** Given $a$ as the length of the side of the cube, find the
distance between the two non-intersecting diagonals of its adjacent
faces.

**Solution 1.**

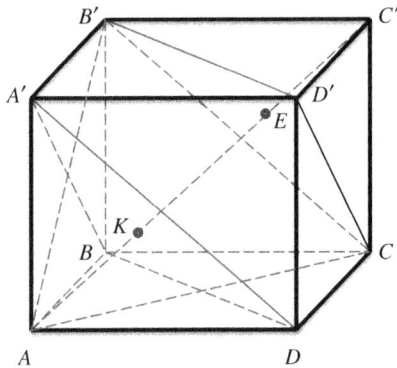

Given cube $ABCDA'B'C'D'$, we need to find the distance between
$BD$ and $D'C$. The diagonals of the faces $ABCD$ and $DCC'D'$, $BD$
and $D'C$, are skew lines. To find the distance between these straight
lines, we may either find the distance between the two parallel planes
containing each of these lines or find the length of the common perpendicular to each line. Let's take the first approach and consider the
two parallel planes $(A'BD)$ and $(B'D'C)$. They are indeed parallel
because $BD \parallel B'D'$ and $A'B \parallel D'C$ as the respective diagonals of the
opposite faces of the cube. So, we see that there are two intersecting
straight lines in one plane, respectively, parallel to the intersecting
two straight lines in the second plane. Therefore, the planes are parallel. If we manage to prove that the diagonal of the cube $AC'$ is
perpendicular to each of these two parallel planes, then the problem will be reduced to finding the distance between the points of
intersection $K$ and $E$ of $AC'$ with each plane. Such a problem looks

manageable. So, let's stick to our plan and prove first that indeed $AC'$ is perpendicular to one of the two parallel planes, for example, to the plane $A'BD$. We see that in our cube, $AB'$ is the projection of $AC'$ onto the plane $A'BB'$. $A'B \perp AB'$ as the diagonals of a square. Therefore, by the theorem of three perpendiculars, $AC' \perp A'B$. Similarly, we can show that $AC' \perp BD$. Hence, we obtained that $AC'$ is perpendicular to the two intersecting straight lines in the plane $A'BD$. It follows that $AC'$ is perpendicular to the plane $A'BD$. Since the planes $(A'BD)$ and $(B'D'C)$ are parallel, then $AC'$ is perpendicular to the plane $(B'D'C)$ as well.

Now, our goal is to find the distance from $C'$ to $(B'D'C)$. This distance equals the length of the altitude $C'E$ in the tetrahedron $C'B'D'C$ in which the base is the regular triangle $B'CD'$ (all three sides are congruent as the diagonals of the congruent squares, $B'C = CD' = B'D' = a\sqrt{2}$) and the other faces are congruent isosceles triangles with the congruent sides $B'C' = CC' = D'C' = a$ (as the edges of the given cube).

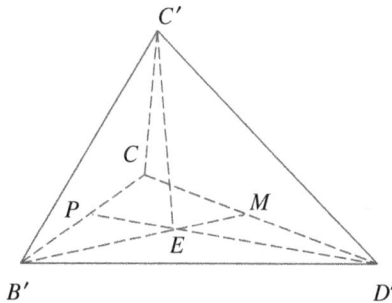

Point $E$, the foot of the altitude $C'E$ of our tetrahedron, is the center of the regular triangle $B'CD'$. I hope the readers can easily justify this fact. $E$ is the point of intersection of the medians $D'P$ and $B'M$ (they are the altitudes and the angle bisectors at the same time) in the regular triangle $CB'D'$. By the Pythagorean Theorem, in the right triangle $D'MB'(\angle M = 90°)$, we find

$$B'M = \sqrt{B'D'^2 - D'M^2} = \sqrt{2a^2 - \frac{2a^2}{4}} = \frac{a\sqrt{3}}{\sqrt{2}}.$$

Consider now the right triangle $C'EB'$ ($\angle E = 90°$). We know that $B'C' = a$ and $B'E = \frac{2}{3} \cdot B'M = \frac{2}{3} \cdot \frac{a\sqrt{3}}{\sqrt{2}} = \frac{a\sqrt{2}}{\sqrt{3}}$ (the medians trisect each other). Therefore, by the Pythagorean Theorem, $C'E = \sqrt{B'C'^2 - B'E^2} = \sqrt{a^2 - (\frac{a\sqrt{2}}{\sqrt{3}})^2} = \frac{a}{\sqrt{3}}$. We will get exactly the same result considering the tetrahedron $AA'BD$ and finding its altitude $AK$, $AK = C'E = \frac{a}{\sqrt{3}}$.

Now, going back to our cube, we will find the length of its diagonal $AC'$. In the right triangle $ACC'$ ($\angle C = 90°$), $AC = a\sqrt{2}$ as the diagonal of the square with the side $a$, and $CC' = a$. Hence, by the Pythagorean Theorem, $AC' = \sqrt{a^2 + 2a^2} = a\sqrt{3}$.

Finally, we can find $KE$, the distance between our two parallel planes containing $D'C$ and $BD$, as the difference between $AC'$ and the sum of $AK$ and $C'E$: $EK = AC' - (AK + C'E) = a\sqrt{3} - 2 \cdot \frac{a}{\sqrt{3}} = \frac{a}{\sqrt{3}}$. So, the distance between $D'C$ and $BD$ is equal to one-third of the diagonal of the cube $AC'$. Indeed, since $AC' = a\sqrt{3}$, then $\frac{1}{3} \cdot a\sqrt{3} = \frac{a}{\sqrt{3}}$, which is the result we obtained above.

**Solution 2.**

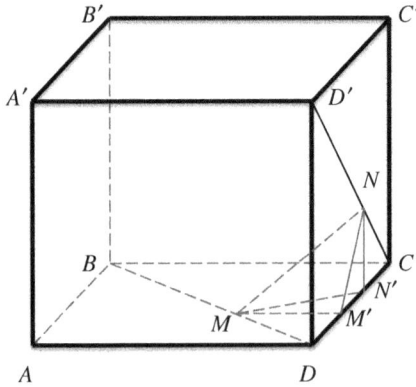

Here, we will take a different path and will try to find the length of the common perpendicular to $D'C$ and $BD$. It does not look like a simple geometrical problem to construct such a common perpendicular. It's not even clear how to approach such a construction. But, bearing in mind the result we got in our Solution 1, we can try using it, and see if it would help us to proceed with the alternative solution. We figured that the distance between the non-intersecting diagonals of the adjacent faces equals one-third of the diagonal of the cube. Thinking about the common perpendicular to $D'C$ and $BD$, we may reasonably assume that it has to be located in a plane perpendicular to each of these diagonals and dividing each in the same ratio as the diagonal of the cube was divided by the two parallel planes containing our non-intersecting diagonals of the adjacent faces. Let's see if our assumption is valid. Denote by $M$ the point on $BD$, such that $MD = \frac{1}{3}BD$ and by $N$ the point on $D'C$, such that $NC = \frac{1}{3}D'C$. Draw $MM' \perp DC$ and $NN' \perp DC$. Then, clearly, $N'C = \frac{1}{3}DC$ and $DM' = \frac{1}{3}DC$. It follows that $M'N' = DM' = N'C = \frac{1}{3}DC$. We obtained that $NN'$ is the altitude and the median at the same time in the triangle $M'NC$. Therefore, $M'NC$ is an isosceles triangle with $M'N = NC$. We also know that $\angle NCD = 45°$ as the angle between the diagonal and the side of a square. So, in our isosceles triangle $M'NC$, both angles by its base equal $45°$, $\angle NCM' = \angle NM'C = 45°$, which implies that the third angle in this triangle is the right angle, $\angle M'NC = 90°$. Hence, $M'N \perp D'C$. Observe now that by construction, $MM' \perp DC$ and $MM' \perp DD'$ since $DD'$ being perpendicular to the plane $ABCD$ is perpendicular to any straight line in that plane; $MM'$ belongs to the plane $ABCD$, so $DD' \perp MM'$. We see that $MM'$ is perpendicular to the two intersecting straight lines in the plane $DCD'$; therefore, it is perpendicular to this plane as well. It implies that $M'N$ is the orthogonal projection of $MN$ onto the plane $DCD\prime$. We just proved that $M'N \perp D'C$, therefore $MN \perp D'C$ as well. In the same way, we can prove that $MN \perp BD$, which will conclude the proof that $MN$ is the sought-after common perpendicular to $D'C$ and $BD$.

The last step will be to calculate the length of $MN$. This can be done, for example, by applying the Pythagorean Theorem to the right triangle $MM'N$ ($\angle MM'N = 90°$). In that triangle, we have $MM' = \frac{1}{3}a$ and $M'N = NC = \frac{1}{3}D'C = \frac{1}{3}a\sqrt{2}$ (one-third of the diagonal of a square). So, finally, we get that

$$MN = \sqrt{(\tfrac{1}{3}a)^2 + (\tfrac{1}{3}a\sqrt{2})^2} = \frac{a}{\sqrt{3}}.$$ This is the same result, as we derived in our Solution 1.

**Solution 3.**

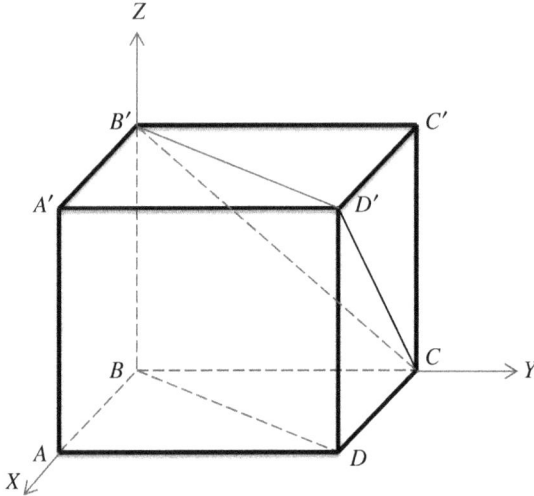

In this solution, we will introduce the Cartesian coordinates in a three-dimensional space, in such a way that $B$ coincides with the origin, and the sides of our cube $BA$, $BC$, and $BB'$ are on the axes $X$, $Y$, and $Z$, respectively. Then, the vertices of the cube can be presented with their coordinates as follows: $B(0,0,0)$, $B'(0,0,a)$, $C(0,a,0)$, and $D'(a,a,a)$.

We know that $BD \parallel B'D'$. It implies that $BD$ is parallel to the plane $B'CD'$ which contains $B'D'$. The distance between $BD$ and $D'C$, therefore, can be found as the distance between $BD$ (any point on $BD$) and the plane $B'CD'$. So, basically our problem here is reduced to finding the distance between any point on $BD$ and the plane $B'CD'$ by applying the Cartesian coordinates techniques.

In a three-dimensional space, the general equation of a plane is written as

$$mx + ny + qz + p = 0.$$

The distance from a random point $M(x_1, y_1, z_1)$ to a plane defined by the equation above is calculated by the formula

$$\frac{|mx_1 + ny_1 + qz_1 + p|}{\sqrt{m^2 + n^2 + q^2}}.$$

First, we need to determine the equation of the plane $B'CD'$. Since each of the points $B'$, $C$, and $D'$ belongs to this plane, the coordinates of each must satisfy the equation of the plane, and we obtain the following system of the linear equations:

$$\begin{cases} m \cdot 0 + n \cdot 0 + q \cdot a + p = 0, & \text{because } B' \in (B'CD') \\ m \cdot a + n \cdot a + q \cdot a + p = 0, & \text{because } D' \in (B'CD') \\ m \cdot 0 + n \cdot a + q \cdot 0 + p = 0, & \text{because } C \in (B'CD'). \end{cases}$$

This system is simplified to

$$\begin{cases} q = -\frac{p}{a}, \\ m = \frac{p}{a}, \\ n = -\frac{p}{a}. \end{cases}$$

Hence, the equation of the plane $B'CD'$ can be written as $\frac{p}{a}x - \frac{p}{a}y - \frac{p}{a}z + p = 0$ or equivalently, $x - y - z + a = 0$. Now, we can find the distance from any point on $BD$ to this plane. For convenience, we can pick point $B(0,0,0)$. Then, the sought-after distance will be calculated as $\frac{|0+0+0+a|}{\sqrt{1^2+1^2+1^2}} = \frac{a}{\sqrt{3}}$.

Determining the distance between two non-intersecting diagonals of the adjacent faces in a cube is more complicated than it might appear at the first glance, but it does make great practice for understanding the principles of solid geometry. By considering different approaches and finding the alternative solutions to this problem, we covered lots of important properties and relations between the lines and planes in a three-dimensional space. We also demonstrated the benefits of applying the Cartesian coordinates method in getting an elegant and relatively simple solution compared to the conventional "pure geometrical" techniques.

In our second solution, we intuitively "predicted" construction steps after locating the points dividing each diagonal of the adjacent faces in a specific ratio. Our prognosis happened to be correct and we managed to get the desired result. However, the drawback of the intelligent guess method is that it does not fully explain why it worked well for us, and it does not disclose the reasons behind the selected path for a solution. In Chapter 9, we discussed how important it is to pose questions about possible generalization of the obtained results. It is natural to inquire if we can get a reasonable clarification, or in other words, if we can solve a more general problem, which would shed light on the construction of the common perpendicular to the two skew lines located in the perpendicular planes. Such a problem can be stated as the following:

> *Given a right dihedral angle between two planes $\delta$ and $\sigma$ that have straight line $AB$ as the common edge, two rays $AP$ and $BQ$ are drawn in $\delta$ and $\sigma$, respectively, and $\angle PAB = \alpha \leq 90°$, $\angle QBA = \beta \leq 90°$. Construct the common perpendicular to the straight lines $AP$ and $BQ$.*

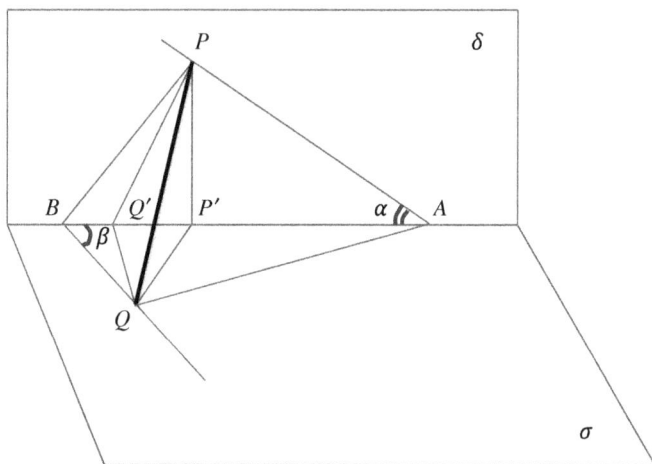

We have two perpendicular planes $\delta$ and $\sigma$ intersecting at $AB$, $P \in \delta$, $Q \in \sigma$. Assume that $PQ$ is the desired common perpendicular to $AP$ and $BQ$. Let's draw $PP' \perp AB$ and $QQ' \perp AB$, and see if we can find a relationship connecting $AP'$ and $AB$,  and $BQ'$ and $AB$. This will allow us to identify (specifically locate) points $P'$ and $Q'$ on $AB$, i.e., explain the construction utilized in our Solution 2 for any general case.

Since $\delta \perp \sigma$ and each of $PP'$ and $QQ'$ is perpendicular to $AB$, the common edge of the two perpendicular planes $\delta$ and $\sigma$, then $PP' \perp \sigma$ and $QQ' \perp \delta$. Assuming that $PQ$ is the desired common perpendicular to $AP$ and $BQ$, we obtain that $Q'P$ is perpendicular to $PA$ as orthogonal projection of $QP$ onto plane $\delta$ and $P'Q$ is perpendicular to $BQ$ as orthogonal projection of $PQ$ onto plane $\sigma$, $Q'P \perp PA$ and $P'Q \perp BQ$.

Consider now the right triangle $PP'A$ $(\angle P' = 90°)$. Denote $AP' = x$. Then,

$$PP' = x \cdot \tan \alpha. \tag{1}$$

We proved that $Q'P \perp PA$; therefore, $\angle Q'PA = 90°$ and considering the right triangle $Q'PA$, we get that $PP'$ is its altitude dropped to the hypotenuse. Hence, $\angle Q'PP' = \angle PAP' = \alpha$.

In the right triangle $Q'P'P$ $(\angle P' = 90°)$, we can now find $Q'P' = PP' \cdot \tan \alpha$ and substituting the value of $PP'$ from (1), we obtain that

$$Q'P' = x \cdot \tan \alpha \cdot \tan \alpha = x \cdot \tan^2 \alpha. \tag{2}$$

Similarly, considering the right triangle $P'QB$ $(\angle Q = 90°)$ and its altitude $QQ'$ to the hypotenuse $BP'$, we get that $\angle Q'QP' = \angle QBQ' = \beta$. Therefore, using (2), we see that in the right triangle $QQ'P'$, $QQ' = Q'P' \cdot \cot \beta = x \cdot \tan^2 \alpha \cdot \cot \beta$ and in the right triangle $BQ'Q$, we have $BQ' = QQ' \cdot \cot \beta = (x \cdot \tan^2 \alpha \cdot \cot \beta) \cdot \cot \beta = x \cdot \tan^2 \alpha \cdot \cot^2 \beta$.

Now, we can express $AB$ as $AB = BQ' + Q'P' + P'A$ and substituting the respective values, obtain that $AB = x \cdot \tan^2 \alpha \cdot \cot^2 \beta + x \cdot \tan^2 \alpha + x$. Using the *trigonometric Pythagorean Identity* $\cos^2 \beta + \sin^2 \beta = 1$ and the definitions of trigonometric functions tangent and cotangent as $\tan \alpha = \frac{\sin \alpha}{\cos \alpha}$ and $\cot \alpha = \frac{1}{\tan \alpha}$, we modify the right side of the last equality as the following:

$$x \cdot \tan^2 \alpha \cdot \cot^2 \beta + x \cdot \tan^2 \alpha + x = x(\tan^2 \alpha \cdot \cot^2 \beta + \tan^2 \alpha + 1)$$

$$= x(\tan^2 \alpha (\cot^2 \beta + 1) + 1) = x \left( \tan^2 \alpha \left( \frac{\cos^2 \beta}{\sin^2 \beta} + 1 \right) + 1 \right)$$

$$= x \left( \tan^2 \alpha \left( \frac{\cos^2 \beta + \sin^2 \beta}{\sin^2 \beta} \right) + 1 \right)$$

$$= x \left( \tan^2 \alpha \cdot \frac{1}{\sin^2 \beta} + 1 \right).$$

So,

$$\frac{AP'}{AB} = \frac{x}{x \left( \tan^2 \alpha \cdot \frac{1}{\sin^2 \beta} + 1 \right)} = \frac{1}{\frac{\tan^2 \alpha}{\sin^2 \beta} + 1} = \frac{\sin^2 \beta}{\tan^2 \alpha + \sin^2 \beta} \tag{3}$$

and similarly,

$$\frac{BQ'}{AB} = \frac{x \cdot \tan^2 \alpha \cdot \cot^2 \beta}{x \left( \tan^2 \alpha \cdot \frac{1}{\sin^2 \beta} + 1 \right)} = \frac{\tan^2 \alpha \cdot \cot^2 \beta}{\tan^2 \alpha \cdot \frac{1}{\sin^2 \beta} + 1}$$

$$= \frac{\tan^2 \alpha \cdot \cot^2 \beta \cdot \sin^2 \beta}{\tan^2 \alpha + \sin^2 \beta} = \frac{\tan^2 \alpha \cdot \cos^2 \beta}{\tan^2 \alpha + \sin^2 \beta}$$

$$= \frac{\frac{\sin^2 \alpha}{\cos^2 \alpha} \cdot \cos^2 \beta}{\frac{\sin^2 \alpha}{\cos^2 \alpha} + \sin^2 \beta} = \frac{\sin^2 \alpha \cdot \cos^2 \beta}{\sin^2 \alpha + \cos^2 \alpha \cdot \sin^2 \beta}.$$

Dividing the numerator and denominator of the last fraction by $(\sin^2 \alpha \cdot \cos^2 \beta)$ and observing that $\frac{1}{\cos^2 \beta} = \tan^2 \beta + 1$, we can modify the last expression as

$$\frac{1}{\frac{1}{\cos^2 \beta} + \frac{\cos^2 \alpha \cdot \sin^2 \beta}{\sin^2 \alpha \cdot \cos^2 \beta}} = \frac{1}{\tan^2 \beta + 1 + \cot^2 \alpha \cdot \tan^2 \beta}$$

$$= \frac{1}{\tan^2 \beta (1 + \cot^2 \alpha) + 1}$$

$$= \frac{1}{\tan^2 \beta \cdot \frac{\cos^2 \alpha + \sin^2 \alpha}{\sin^2 \alpha} + 1} = \frac{1}{\tan^2 \beta \cdot \frac{1}{\sin^2 \alpha} + 1}$$

$$= \frac{\sin^2 \alpha}{\tan^2 \beta + \sin^2 \alpha}.$$

So, finally,

$$\frac{BQ'}{AB} = \frac{\sin^2 \alpha}{\tan^2 \beta + \sin^2 \alpha}. \tag{4}$$

With the given angles $\alpha$ and $\beta$, formulas (3) and (4) allow finding a specific location of each point $P'$ and $Q'$ on $AB$, which in turn would allow constructing $PQ$ as the common perpendicular to skew straight lines $AP$ and $BQ$.

Going back to our Problem 9, we know that $\alpha = \beta = 45°$. So, applying these formulas in our second solution, we get $\frac{MD}{BD} = \frac{NC}{D'C} = \frac{DM'}{DC} = \frac{\sin^2 45°}{\tan^2 45° + \sin^2 45°} = \frac{\frac{1}{2}}{1 + \frac{1}{2}} = \frac{1}{3}$. We see that now there is no need for any intelligent guesses; we got a very specific explanation of how to select points $M$ on $BD$ and $N$ on $D'C$ to draw the common perpendicular to lines $BD$ and $D'C$.

By determining formulas (3) and (4), we arrived at an important generalization of our Problem 9. Both general formulas prove useful in any similar problems.

Furthermore, we can also define the general formula to calculate the distance between two skew lines. Recall that in any right triangle, the altitude dropped to the hypotenuse divides it into two segments such that the length of the altitude equals the geometrical mean of these two segments. Then, in the right triangle $Q'PA$, $P'P^2 = AP' \cdot P'Q'$. Also observe that the right triangles $QQ'P'$ and $BQP'$ are similar, $\triangle QQ'P' \sim \triangle BQP'$, because they have equal angles. Therefore, $\frac{P'Q}{BP'} = \frac{P'Q'}{P'Q}$, from which $P'Q^2 = BP' \cdot P'Q'$.

Applying the Pythagorean Theorem to the right triangle $PP'Q(\angle P' = 90°)$ and substituting the respective values from the above expressions, we obtain that

$$PQ^2 = P'P^2 + P'Q^2 = AP' \cdot P'Q' + BP' \cdot P'Q'$$
$$= P'Q' \cdot (AP' + BP') = P'Q' \cdot AB.$$

From the last expression, it implies that $PQ = \sqrt{P'Q' \cdot AB}$.

Knowing the location of points $P'$ and $Q'$, we can easily calculate the length of the segment $P'Q'$ as $P'Q' = AB - AP' - BQ'$. Therefore, the obtained formula $PQ = \sqrt{P'Q' \cdot AB}$ indeed allows calculating the desired distance between two skew lines under the given conditions.

The above solution of the general case is not unique. We suggest the readers seek for the alternative solutions, for example, by applying the vector algebra techniques.

The next problem requires more advanced mathematical knowledge and some ingenuity in applying the learned concepts and techniques.

**Problem 10.** Given a straight line passing through point $A(-3, 1)$ tangent to the graph of the function $y = \sqrt{8 - x^2}$, find the angle it forms with the $X$-axis.

**Solution 1.** To find the angle formed by the given tangent to the graph of the given function, we have to determine the slope of the tangent. Recall that the geometrical meaning of the slope of a straight line is the tangent of an angle it forms with $X$-axes. So, by finding the slope, we would be able to determine the angle in question as well.

The general equation of a straight line in the *point-slope form* is written as

$$y - y_1 = m(x - x_1).$$

Since $A$ lies on the line, its coordinates satisfy the equation. Hence, it implies that the equation of the given tangent to the graph of the given function is $y - 1 = m(x + 3)$, or equivalently, $y = mx + 3m + 1$, where $m$ is the slope of the tangent line.

Let $M(x_M, y_M)$ be the point of tangency of the given straight line and the graph of the given function. The coordinates of $M$ must satisfy the following equations of the straight line and the function (system of equations)

$$\begin{cases} y_M = mx_M + 3m + 1, \\ y_M = \sqrt{8 - x_M^2}. \end{cases} \qquad (*)$$

Setting these expressions equal to each other, we obtain

$$mx_M + 3m + 1 = \sqrt{8 - x_M^2}.$$

Squaring both sides gives

$$8 - x_M^2 = m^2 x_M^2 + 9m^2 + 1 + 6m^2 x_M + 6m + 2mx_M,$$

which can be rewritten as

$$x_M^2(m^2 + 1) + 2mx_M(3m + 1) + 9m^2 + 6m - 7 = 0.$$

The given straight line will be tangent to the graph of the given function when the above equation has only one solution. Considering this equation as quadratic for $x_M$, it implies that the discriminant of the equation has to equal 0. Let's find it (to simplify calculations, we will be looking for $\frac{D}{4}$ rather than $D$):

$$\frac{D}{4} = m^2(3m + 1)^2 - (m^2 + 1)(9m^2 + 6m - 7).$$

This simplifies to $\frac{D}{4} = -m^2 - 6m + 7$. Letting $\frac{D}{4} = 0$, we obtain the quadratic equation for $m$, $m^2 + 6m - 7 = 0$, solutions of which are

$m_1 = 1$ and $m_2 = -7$. Substituting the found values back into the system (*), we can verify that 1 satisfies it, but $-7$ does not and has to be rejected. We leave this verification for the readers to do. Therefore, the slope equals 1. So, we get that $\tan \alpha = 1$, which implies that the desired angle equals $45°$, $\alpha = 45°$, because $\tan 45° = 1$.

**Solution 2.**

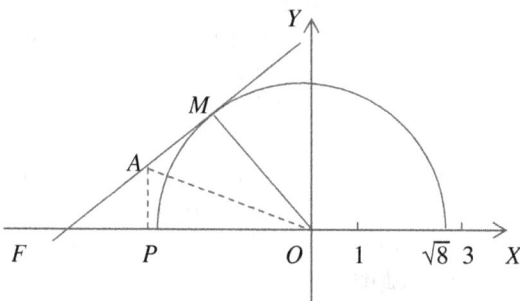

The graph of the function $y = \sqrt{8 - x^2}$ is the semicircle with the center $O$ in the origin and with the radius equal to $\sqrt{8}$. Let $FM$ be the given tangent to the semicircle at point $M$ containing the given point $A(-3, 1)$. Then, it has to be perpendicular to the radius drawn from $O$ to $M$, $FM \perp OM$. Now, drop $AP$ perpendicular to $X$-axes. Knowing the coordinates of $A(-3, 1)$, we see that $OP = 3$, $AP = 1$, and applying the Pythagorean Theorem to the right triangle $APO$ gives $AO = \sqrt{9 + 1} = \sqrt{10}$.

From the right triangle $AMO$, $\sin \angle OAM = \frac{OM}{OA} = \frac{\sqrt{8}}{\sqrt{10}}$. Therefore, using the trigonometric *Pythagorean Identity* $\sin^2 \alpha + \cos^2 \alpha = 1$ and the definition of tangent of an angle as $\tan \alpha = \frac{\sin \alpha}{\cos \alpha}$, we get that

$$\tan \angle OAM = \frac{\frac{\sqrt{8}}{\sqrt{10}}}{\sqrt{1 - \frac{8}{10}}} = \frac{\sqrt{8}}{\sqrt{2}} = 2.$$

From the right triangle $APO$, $\tan \angle AOP = \frac{AP}{PO} = \frac{1}{3}$.

Since angle $OAM$ is the exterior angle of the triangle $OAF$, $\angle OAM = \angle AFO + \angle AOF$, from which $\angle AFO = \angle OAM - \angle AOF$. It follows that $\tan \angle AFO = \tan(\angle OAM - \angle AOF)$.

Noticing that $\angle AOF = \angle AOP$ and applying the identity $\tan(x - y) = \frac{\tan x - \tan y}{1 + \tan x \tan y}$, we obtain that $\tan \angle AFO = \frac{2 - \frac{1}{3}}{1 + 2 \cdot \frac{1}{3}} = \frac{5}{5} = 1$. Hence, $\angle AFO = 45°$.

The considered problem perhaps is more complicated than any other problem in the book, and while looking for the alternative solutions to this problem, we covered a wide spectrum of topics in analytical geometry, algebra, plane geometry, and trigonometry. We demonstrated also how useful trigonometric identities are in connecting algebraic and geometric issues. In spite of its difficulty, we believe it was worthwhile to make an effort in tackling this problem. The examined solutions allowed for a deeper understanding of the mathematical concepts involved and provided the opportunities for practicing various useful techniques.

To conclude this chapter, we offer for the readers two more interesting problems. The first is about calculating the length of a segment parallel to the bases of the given trapezoid and dividing it into two trapezoids of equal areas. Try to find several alternative solutions to this problem; two such solutions we offer in our "Solutions to Problems" section at the end of the book. The second problem was formulated by the prominent English mathematician John Wallis (1616–1703), and it is about finding a rectangle with the maximum area out of rectangles with the same perimeter.

**Problem 11.** In a trapezoid $ABCD$ *(AB ∥ CD)*, $AB = a$, and $CD = b$. $MN$ is parallel to the bases of the trapezoid and divides it into two trapezoids of equal areas. Find the length of $MN$.

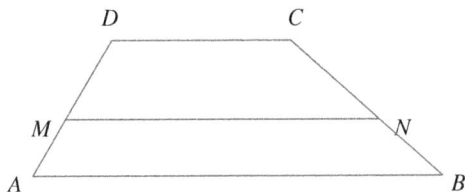

**Problem 12.** John Wallis's problem.

Prove that of all rectangles with the same perimeter, a square has the maximum area. Show geometrical and algebraic proof.

# Chapter 11

# Eureka!

The purpose of this book is to ease and elucidate the process of solving problems. It is an overview not only of the steps, but also the mental operations that are useful in the process as well. I strongly believe that seriously studying mathematics and learning how to solve problems help individuals prepare themselves to face multiple challenges in their lives. While discussing and solving non-routine problems during the course of the book, we witnessed on multiple occasions how important a bright idea can be to inspire a path for a problem's solution. In some circumstances, while having very limited time, it becomes critical to promptly understand, evaluate, and come up with a great idea for solving a particular problem. I noticed that when reading Dan Brown's books, at some moments, I felt uncomfortable; I would even say, tense and edgy. His main character, professor Langdon, is often brought into extreme situations when he has to solve uneasy problems and find the answers to difficult questions in a matter of a few minutes. As you read those interesting stories, you envision yourself in professor Langdon's shoes and try to answer those same questions while having the same restrictions as him. This is not an easy task! But, sometimes, we face such similar occasions in our lives and have to survive by finding the proper solutions. The tasks and problems in nature are different just as their respective surrounding circumstances and facts, but ultimately, they all require a deep understanding of the subject, the ability to make proper decisions based on the knowledge and experience, creativity, ingenuity, and surely, logical thinking. In some cases, the ability to solve a problem under time pressure becomes critical.

One of the extreme examples is presented by the 1986 Chernobyl power plant explosion. A 2019 HBO television miniseries depicted the Chernobyl nuclear power plant disaster and the events of the cleanup effort that followed. Being a Kiev resident since my birth, I well remember those dramatic days of late April and beginning of May of 1986. We did not know the entire truth about the magnitude of the catastrophe because the Soviet government concealed lots of information about the Chernobyl tragedy. Only recently, many facts have been finally revealed. A world-famous scientist academician Valery Legasov (1936–1988), one of the members of the government commission, took the most important decisions in fire contamination and analysis of how to avoid repeated accidents. He and his associates in a very limited time under extremely dangerous life-threatening conditions had to solve problems that nobody in human history had ever faced before. They managed to make proper decisions and implement them to save millions of lives. It was under Legasov's command that it was decided to dump boron carbide in large quantities from helicopters to act as a neutron absorber and prevent any renewed chain reaction. Lead was included as a radiation absorber, as well as sand and clay, which the scientists hoped would prevent the release of particulates. Later, steps were taken to prevent melted radioactive material from reaching the water in the lower cooling system, so a tunnel was built to prevent radioactive elements from reaching groundwater.

It's hard to say what step in the problem-solving process is more important. It is impossible to solve any problem without good knowledge of the subject matter, a clean understanding of the given information and the question that is being asked. By observing, practicing, and by thinking out of the box, you embark toward achieving your goal. The idea of where to start, in what direction to go, and what steps to take, may not emerge right away. It may come to you piece by piece, especially after some failed attempts and trials. But, what a rewarding experience it will be after you overcome a difficult problem and get the final result! When a bright idea hits you, it enlightens the entire picture and you feel a great deal of self-satisfaction and self-confidence. All your excitement can be best expressed in this one famous Archimedes' quote "Eureka!" Ancient Greek mathematician and inventor Archimedes (c. 287 B.C.–c. 212 B.C.) yelled "Eureka!" after he had stepped into a bath and noticed that the water level rose; he suddenly understood that the volume of water displaced must be equal to the volume of the part of his body he had submerged. This discovery allowed him to solve a previously intractable problem of measuring the volume of irregular objects with precision. Since then, the phrase "Eureka!" has been a synonym for a bright idea, or in our case a discovery leading to a breakthrough in solving a problem. In the last chapter of the book, we put together a few more "Eureka-type" challenges for your enjoyment. Originality and creativity are crucial in solving the problems offered below. We hope the readers accept the challenge and will have fun and success at arriving to great ideas for pathfindings in the final labyrinths in the book.

## 11.1 Weighing Problems

**Problem 1.** Out of nine coins of the same denomination, eight weigh the same, and one coin is lighter than the others. Find the counterfeit in two weighings on a pan balance scale, without using weights. Can you extend this problem and find one counterfeit among 27 coins with three weighings only?

**Problem 2.** There are five scale weights of 1,000 grams, 1,001 grams, 1,002 grams, 1,004 grams, and 1,007 grams. There is no signed weight on any of them and they all look the same. You can use the weighing scale machine that shows the weight in grams. Can you identify the weight of 1,000 grams by only using three weighings?

**Problem 3.** Four bank robbers successfully shut down an alarm and got to the bank safe. They managed to open the safe and retrieved nine bags with gold coins of the same denomination. The bank insider told them that all the coins in one of the bags are counterfeit and have no value. One genuine coin weighs 10 grams and a counterfeit coin is lighter than a genuine coin by 1 gram. The robbers have very limited time to escape and have to decide which eight bags to take (they have only four pairs of hands to carry the bags). There is a balance weight scale with weights next to the safe. Is it possible with just one weighing (they don't even have enough time for two weighings) to identify the bag with the counterfeits?

**Problem 4.** In a small grocery store, the owner has a weighing balance scale with its shoulders having different lengths. The first time he weighed 1 pound of strawberries using the left shoulder. When the same customer asked to add another pound of strawberries, the owner fulfilled the order, but used the right shoulder to weigh strawberries. Did the customer get exactly 2 pounds of strawberries, less or more?

**Problem 5.** There are four identical coins one of which is counterfeit. The weight of a genuine coin is 5 grams. The weight of a counterfeit coin is different but it is unknown whether it is heavier or lighter than 5 grams. Is it possible with only two weighings on a pan balance scale to identify the counterfeit and determine whether it is heavier or lighter than a genuine coin if you have as a supplemental tool a 5-gram scale weight?

**Problem 6.** There are 101 identical coins one of which is counterfeit. Is it possible with only two weighings on a pan balance scale to determine whether the counterfeit coin is heavier or lighter than a genuine coin?

## 11.2   Playing with Numbers

**Problem 7.** Replace letters and stars (can be any digit) with numbers:

a)
$$\begin{array}{r} a \\ + bb \\ \underline{a} \\ ccc \end{array}$$

b)
$$\begin{array}{r} *\,* \\ \times\ \underline{*\,*} \\ *\,* \\ \underline{*7} \\ *\,*\,*\,* \end{array}$$

c)
$$\begin{array}{r} *\,*\,* \\ \times\ \underline{*\,8} \\ *\,*\,* \\ \underline{*\,*\,*5} \\ *\,*\,*\,*0 \end{array}$$

d) Replace letters and stars with numbers. It is an interesting cryptogram expressing in three different languages the same product *two times two equals four*:

| In Russian: | In Polish: | In Spanish: |
|---|---|---|
| $два$ | $dwa$ | $dos$ |
| $\times\ \underline{два}$ | $\times\ \underline{dwa}$ | $\times\ \underline{dos}$ |
| $****$ | $**^{r}*$ | $****$ |
| $***^{6}$ | $****$ | $***\,s$ |
| $\underline{^{e}***}$ | $\underline{y***}$ | $\underline{****}$ |
| $четыре$ | $cztery$ | $cuatro$ |

**Problem 8.** Evaluate without a calculator
$1987 \cdot 19861986 - 1986 \cdot 19871987$.

**Problem 9.** Evaluate without a calculator $\frac{1}{2} + \frac{1}{2\cdot3} + \frac{1}{3\cdot4} + \cdots + \frac{1}{99\cdot100}$.

**Problem 10.** Evaluate without a calculator
$\frac{1}{\sqrt{1}+\sqrt{2}} + \frac{1}{\sqrt{2}+\sqrt{3}} + \frac{1}{\sqrt{3}+\sqrt{4}} + \cdots + \frac{1}{\sqrt{99}+\sqrt{100}}$.

**Problem 11.** In the expression 123456789, insert "plus" and "minus" signs between certain digits so the calculated value of the obtained new expression equals 100. Can you show several alternatives?

**Problem 12.** With 17 matches, form the following "equality" as written below:

$$\frac{XXIII}{VII} = II$$

The equality shows that $23 : 7 = 2$. Change the position of one of the matches so the equality holds with the precision of 0.002.

**Problem 13.** Is it possible to divide twelve by two to get seven?

**Problem 14.** Is it possible to add four digits to the right of 9999 so the obtained eight-digit number becomes a perfect square of some natural number?

**Problem 15.** (Cryptography Problem) Decode the following cryptogram for the difference of positive numbers.

$$
\begin{array}{r}
\text{ROME} \\
- \ \ \underline{\text{SUM}} \\
\text{RUSE}
\end{array}
$$

Solving this puzzle, keep in mind that generally the cryptograms have a hint for decoding in their presentation.

**Problem 16.** Is a number $\underbrace{111\cdots1}_{81\ digits}$ divisible by 81?

**Problem 17.** Is it possible to factor out $(2^{1982}+1)$ into two natural factors each of which is not less than 1000?

**Problem 18.** Prove that number $1\underbrace{000\cdots0}_{1994\ zeros}1$ is a composite number.

## 11.3 Equations to the Rescue

**Problem 19.** The cost of 25 oranges in dollars equals the number of oranges one can buy for \$1. How many oranges can be purchased for \$3?

**Problem 20.** In 2 years, a boy will be twice as old as he was 2 years ago. In 3 years, a girl will be three times as old as she was 3 years ago. Who is older, the boy or the girl?

**Problem 21.** There is a two-digit number that is two times the sum of its digits. What is this number?

**Problem 22.** At what time after the clock showed 2 pm will the minute and the hour hands overlap?

**Problem 23.** The only inhabitants of a mysterious island in Wonderland are the three-headed dragons and 40-legged dragons. There are a total of 26 heads and 298 legs among them. How many legs does each three-headed dragon have if each of 40-legged dragons has one head?

**Problem 24.** There are 10 big size travel bags listed in a store inventory. There are 10 bags of a smaller size inserted in some of them. Some of the smaller size bags also have 10 bags of even smaller size inserted in them. How many bags are there of all three sizes if it is known that there are total of 54 bags with smaller bags that are inside them?

**Problem 25.** The distance between street $A$ and street $C$ is 20 miles. A car departed from $A$ to $C$ at the same time as a pedestrian left his house on street $B$. Street $B$ is located between $A$ and $C$ at a distance of 15 miles from $A$. The pedestrian was walking in the direction to $C$. At the same time, a bus left from $C$ to $B$. In what time did the car overtake the pedestrian, if it is known that it happened in a half-an-hour after meeting of the bus and the car, and the pedestrian spent three times less time walking before meeting with the bus than the car spent driving before it met with the bus?

**Problem 26.** Liana is six times younger than her great-grandfather. If you insert a 0 between the digits composing her age, then you will get her great-grandpa's age. How old is she?

## 11.4 Recreational Geometry — Can You Do It?

**Problem 27.** Cross all four vertices of a given square with three straight lines without taking your pencil off the paper.

•    •

•    •

**Problem 28.** Make a square with nine dots as shown below. Can you cross all the dots with four straight lines without taking your pencil off the paper?

**Problem 29.** Can you draw the picture shown below without taking your pencil off the paper and not passing any line twice?

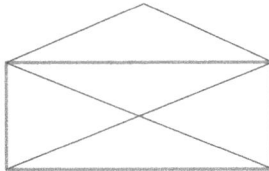

**Problem 30.** Can you cross all the dots on the picture below with straight lines without taking your pencil off the paper?

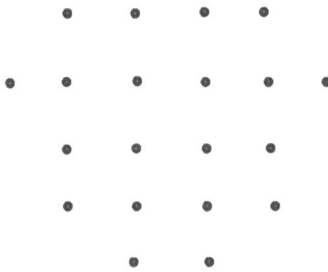

**Problem 31.** Can you arrange 10 points on 5 segments in such a way that there are 4 points on each segment?

**Problem 32.** Can you arrange 10 chairs in a room in such a way that there are 3 chairs by each of the 4 walls?

**Problem 33.** Can you arrange 9 points on 10 segments in such a way that there are 3 points on each segment?

**Problem 34.** Three coins of different nominations from some foreign country are placed on a table in such a way that they touch each other and their centers form a right triangle. Can you identify the minimal possible radii of these coins by expressing the results in natural numbers? To make it a little more challenging, do not use a pencil and paper; solve the problem in your head.

**Problem 35.** Is it possible to draw a straight line cutting a right triangle into two isosceles triangles which have the same area?

**Problem 36.** Cut the cross consisting of five congruent squares in several pieces such that after assembling them, you can get a square.

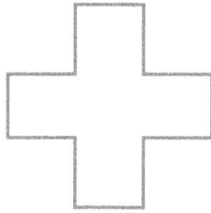

**Problem 37.** The sum of the diagonals of any trapezoid is always greater than the sum of its bases. Is this statement true or false? Justify your response.

**Problem 38.** How should you draw $BM$ and $BK$ (see the picture below), so that the area of the quadrilateral $BMDK$ is $\frac{1}{3}$ of the area of the parallelogram $ABCD$ ?

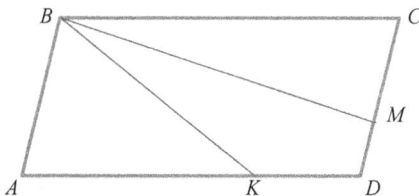

**Problem 39.** In a regular 9-gon, the diagonals cut it in several triangles. Compare the sum of the areas of the shaded triangles and the sum of the areas of the white triangles. Which sum is greater?

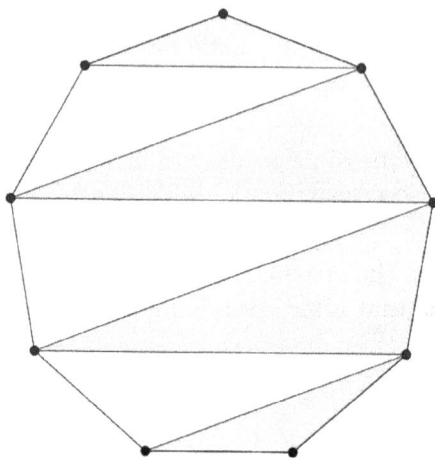

## 11.5   Sense of Humor, Common Sense, and a Little Bit of Logic

**Problem 40.** Two friends Bob and Steve live in a 12-story apartment building; Bob lives on the 8th floor and Steve lives on the 10th floor. Returning back from school, they stepped out on the balcony in each apartment and simultaneously shouted to each other "Hi, I am home!" Who will be the first to hear the friend's greeting? Assume that they yell with the same loudness, there is no wind, and no other physics effects are to be involved here at all.

**Problem 41.** Matthew, who just learned how to play chess, fell in love with the game. After a few days, he announced that he can play with the chess world champion and the runner-up for the title at the same time, and collect at least one point in both games (in a chess game, a point is awarded for a win and half-point for a draw). Can he?

**Problem 42.** While walking in a park, two brothers Alex and Bryan entered a big circle in an alley of trees. Bryan started counting the trees. Alex followed him, but started counting from a different tree, while he was walking in the same direction. The tree that was the 20th for Alex was 7th for Bryan, and the 7th tree for Alex happened to be 94th for Bryan. How many trees were standing on the circle alley?

**Problem 43.** Gary spends $\frac{1}{3}$ of his time studying in college, $\frac{1}{4}$ — playing basketball, $\frac{1}{5}$ — listening to music, $\frac{1}{6}$ — watching TV, and $\frac{1}{7}$ — socializing with friends online. Is it possible for him to live his life like that?

**Problem 44.** Is it a correct statement that there are always two people at any party who have the same number of acquaintances (if A knows B, then B knows A)?

**Problem 45.** Biologists discovered a new type of lily-flower from which, when left alone, its bulblets grow into bulbs and grow new plants next to the parent. These bulblets grow so fast that every day the new lilies are doubled, and eventually, in 10 days the lake is covered entirely by the flowers. Next year, two such lilies were planted on the same lake. In how many days will the flowers cover the entire lake?

**Problem 46.** There are three boxes of the same size with two small balls inside each of them. There are two black balls in the first box, two white balls in the second box, and a black and white ball in the third box. Each box has a label with an indication of what is inside: "Black balls", "White balls", and "White and black". The problem is that all the labels were attached to the wrong boxes. How can you determine what is inside each box by taking out only one ball from any of the boxes?

**Problem 47.** In one month, three Saturdays fall on even dates. What day of a week was the 28th of this month?

**Problem 48.** There are six glasses on the table, the first three with water and the last three empty. Is it possible by touching only one of those glasses to have the glasses arranged in such an order that every glass with water is followed by an empty glass?

**Problem 49.** You have two pieces of Bickford fuse, each of which burns in 1 minute. You want to light your New Year Green tree star in exactly 45 seconds and your clock has no second hand. Can you

use these two pieces of fuse to time 45 seconds and achieve your goal on time? You may not use scissors.

**Problem 50.** There are three light bulbs in a room and three switches located in another room. Each switch can be either "ON" or "OF" and is connected to one bulb. How can you determine which switch is connected to what bulb if you are allowed to enter the room with bulbs only once?

*Hint*: Some basic knowledge in physics should help.

**Problem 51.** Alex said that on the day before yesterday he was 10 years old, and next year he is going to be 13 years old. Is it possible?

**Problem 52.** What is the fastest way to cover the distance between cities $A$ and $B$:

(1) Ride the whole distance on a bicycle

or

(2) Drive the first half of the distance between $A$ and $B$ on a motorcycle with a speed that is double that of the bicycle's speed, and then walk the second half of the way with a speed that is half of the bicycle's speed?

**Problem 53.** The problem was offered by I. Akulich in magazine "Квант" (in Russian), #3, 1999 year.

There are several candies on a table. The boys who were invited to the party took candies from the table in the following manner:

The first boy took $\frac{1}{10}$ of all candies. The second boy took $\frac{1}{10}$ of the remaining candies and $\frac{1}{10}$ of the candies taken by the first boy. The third boy took $\frac{1}{10}$ of the remaining candies and $\frac{1}{10}$ of the candies taken by the first and the second boy. And so on, till all the candies have been taken. How many boys were there and who got the most candies?

**Problem 54.** The problem was offered by I. Akulich in magazine "Квант" (in Russian), #3, 2001 year.

When asked about his age, a problem-posing professor answered, "If you multiply the year in which I was 43 by the year in which I was 45, and then divide by the year in which I was born, then you will get the year...". At that moment, he was interrupted by one of his students, "Sorry, professor, you don't have to continue. I know how to figure out the year in which you were born". Do you believe he can?

**Problem 55.** What should be the next character instead of "?"?

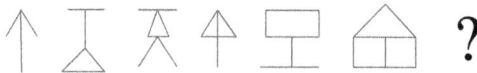

The final challenge in the book is one of my favorite moral-test problems, the solution of which is unique for anybody attempting it. It is really your choice and it is up to you what decision to make. It does take ingenuity and unorthodox thinking to get to the optimal outcome.

## 11.6  Final Problem — Moral Test

Imagine that you are driving your car in a nasty snow storm late at night, and all of a sudden you see three people waiting for a bus at the bus stop:

1. A very old lady that looks sick and is at the verge of passing out;
2. Your old friend who saved your life several years ago;
3. A beautiful lady/gentleman, the perfect image of your lifetime dreams.

You can accommodate just one person from that bus stop. Who would you take as a passenger in your two-seater car?

I consider this problem as an exclamation sign at the end of our journey, and I have no doubt you should agree with me if you manage to solve it saving all three persons (this is the tip; yes, it is possible!).

Don't look at the solution at the end of the book before you make your choice. It is a moral test, first of all, but it is a logical and intriguing problem at the same time, which does have an off-beat solution. This is great evidence of how with some creativity and inventive thinking, we can accomplish incredible results even in extreme situations.

To be a successful problem solver, one needs a combination of natural talent, intuition, knowledge, skill, and constant practice. Tackling a hard problem often causes some degree of frustration and disappointment. However, all your struggles in finding the way out of a math labyrinth are well paid off! By solving difficult and unorthodox problems, one not only learns useful techniques which apply across the spectrum of mathematics, but he/she also gains proficiency and elevates confidence. Our goal was to give some hints and practice in overcoming non-conventional challenges allowing readers to experience mathematical pleasure and the thrill of mathematical discovery. I hope readers find it enjoyable, useful, and helpful; this is the best reward for the author.

> *Mathematics is a more powerful instrument of knowledge than any other that has been bequeathed to us by human agency.*

> — René Descartes

# Solutions and Answers to Problems

**Chapter 2**

**Problem 19.** Our number $N = 1\underbrace{00\ldots00}_{49 \text{ zeros}}2\underbrace{00\ldots00}_{99 \text{ zeros}}1$ has 151 digits. It is not hard to see that $N > \left(10^{50}\right)^3$ because $(10^{50})^3 = 10^{150} = 1\underbrace{00\ldots00}_{150 \text{ zeros}}$.

On the other hand, recalling the formula $(a+b)^3 = a^3 + 3a^2b + 3ab^2 + b^3$, we get that $N = 10^{150} + 2\cdot 10^{100} + 1 < 10^{150} + 3\cdot 10^{100} + 3\cdot 10^{50} + 1 = (10^{50}+1)^3$. Thus, it becomes obvious that

$$(10^{50})^3 < N < (10^{50}+1)^3. \tag{*}$$

Since $10^{50}$ and $10^{50}+1$ are the two consecutive natural numbers, then there is no natural number, the cube of which satisfies (*).

**Problem 20.** Don't be confused by the language of the problem. Follow the money movements and see what actually happened. After giving $40 change to the customer, the sales person had $10 of gross sales to record. Then, he had to return $50 borrowed from his associate to reimburse him for the fake bill, this is the actual loss he had.

**Problem 21.** Noticing the reciprocal fractions in both sides of the equation $\frac{x-49}{50} + \frac{x-50}{49} = \frac{49}{x-50} + \frac{50}{x-49}$, we can rewrite it as $\frac{x-49}{50} - \frac{50}{x-49} = \frac{49}{x-50} - \frac{x-50}{49}$. In other words, the way the equation is originally written provides the hint for the plan on how to solve it.

Modifying the last equation, we get $\frac{(x-49)^2-50^2}{50(x-49)} = \frac{49^2-(x-50)^2}{49(x-50)}$. Applying the formula for the difference of squares in each nominator, the last equation simplifies to $\frac{(x-99)(x+1)}{50(x-49)} = \frac{(99-x)(x-1)}{49(x-50)}$. At this point, we see that the first root of the equation is $x = 99$ and after simplification, we get $\frac{x+1}{50(x-49)} = \frac{1-x}{49(x-50)}$. Solving this equation yields

$$49(x+1)(x-50) = 50(x-49)(1-x);$$
$$99x^2 - (50^2 + 49^2)x = 0;$$
$$x(99x - 50^2 - 49^2) = 0;$$
$$x = 0 \text{ or } x = 49\frac{50}{99}.$$

**Answer:** $0$, $49\frac{50}{99}$, $99$.

**Problem 22.** Nine years ago, the sum of ages of all the family members was 40 and now it is 65. So, for the past 9 years it increased by $65 - 40 = 25$ years. But, if the family consisted of three persons 9 years ago, such a total increase should have been $3 \cdot 9 = 27$ years, not 25 years. It implies that 9 years ago, the son was not born yet, and he has to be $25 - 9 \cdot 2 = 7$ years old now. Then, he was $7 - 4 = 3$ years old 4 years ago, and since the father was nine times as old as the son, his age was $9 \cdot 3 = 27$, 4 years ago. Therefore, the father is $27 + 4 = 31$ years old now.

Another way of solving this problem is through applying equations.

Let $x$ years be the father's age now, $y$ years be the mother's age now, and $z$ years be the son's age now. According to the first condition of the problem,

$$x + y + z = 65. \tag{1}$$

If this family consisted of the three persons 9 years ago, then the sum of ages of all three of them was supposed to be $65 - 9 \cdot 3 = 38$. Since it is given that this sum was 40 years, not 38, we conclude that the son was not born yet at that time 9 years ago. So, the sum of ages of the two family members, father and mother, 9 years ago was

$$(x - 9) + (y - 9) = 40. \tag{2}$$

Finally, we know that 4 years ago, the father was nine times older than his son, which can be written as

$$x - 4 = 9(z - 4). \tag{3}$$

Combining (1)–(3) we get the system of linear equations

$$\begin{cases} x + y + z = 65, \\ (x - 9) + (y - 9) = 40, \\ x - 4 = 9(z - 4). \end{cases}$$

It can be rewritten as

$$\begin{cases} x + y + z = 65, \\ x + y = 58, \\ x = 9z - 32. \end{cases}$$

Substituting the expression for $x$ from the third equation into the first two equations and simplifying gives

$$\begin{cases} y + 10z = 97, \\ y + 9z = 90, \\ x = 9z - 32. \end{cases}$$

Subtracting the second equation from the first gives $z = 7$. Then, $x = 31$, and $y = 27$.

**Answer:** The father is 31 years old now.

**Problem 23.** Since the father arrived at the bus stop 30 minutes early and got back home with his son 10 minutes earlier than usual, then the father must have spent 20 minutes more for the entire trip. Assume he walked $t$ minutes until he met his son's car. Hence, they met at 7 hours and 30 minutes $+ t$ minutes. Since the rest of the way after meeting his son, they were driving in a car, the time the father spent for walking is 20 minutes more in comparison with the time spent for the same distance by driving a car. But, under normal circumstances, the car was supposed to be at the bus station at 8 pm. So, the time needed for the car to cover the distance he was walking is $(30 - t)$ minutes.

We have already established that the car spends 20 minutes less than the walking man for this distance. Hence, the car should cover this distance in $(t-20)$ minutes. We get the equation, $30-t = t-20$, solving which we get $t = 25$ minutes. So, they met at 7.55 pm.

Another way to look at this problem is to do it from the son's perspective. Since they got back home 10 minutes earlier than usual, the son spent 5 min less time in each direction than usual. It means that he met his father not at 8 pm at the bus station as usual, but he met him walking from the station at 7.55 pm (5 minutes earlier). This is perhaps a more elegant and easier solution than by using equations. Sometimes, converting to algebra is not the best choice (still valid though), as soon as your analysis allows for a direct conclusion in getting the desired result.

**Problem 24.** In the equality $\overline{AB} \cdot \overline{DC} = \overline{KUKU}$, we see that he multiplied two two-digit numbers and got a four-digit number in the product, such that its digits are the same on even and odd spots. This is possible when you multiply any two-digit number by 101, i.e., $\overline{KUKU} = \overline{KU} \cdot 101$. For example, $45 \cdot 101 = 4545$. Therefore, we can rewrite the original equality as $\overline{AB} \cdot \overline{DC} = \overline{KU} \cdot 101$. This implies that at least one of the prime factors on the left-hand side, $\overline{AB}$ or $\overline{DC}$, has to be divisible by 101, which is a prime number. This is impossible because a two-digit number cannot be divisible by a three-digit number. So, clearly, he made a mistake replacing digits for letters in his problem.

**Problem 25.** We need to solve the equation

$$(x^2 + 3x - 4)^3 + (2x^2 - 5x + 3)^3 = (3x^2 - 2x - 1)^3.$$

There has to be some trick to be used here. The equation looks difficult, and a straightforward approach to it is bound to fail. As with many problems studied in the book, we have to look closely at the given information. It should not be hard to see that the sum of the bases of powers on the left-hand side equals the base of the power on the right-hand side, i.e., $(x^2+3x-4)+(2x^2-5x+3) = 3x^2-2x-1$. So let's try simplifying the equation introducing new variables:

$$u = x^2 + 3x - 4 \quad \text{and} \quad v = 2x^2 - 5x + 3.$$

Our equation then can be rewritten as

$$u^3 + v^3 = (u + v)^3.$$

Applying the formula $(u + v)^3 = u^3 + v^3 + 3uv(u + v)$, we simplify the equation and get $3uv(u + v) = 0$. It follows that either $uv = 0$ or $u + v = 0$.

So, going back to our substitutions, we have

$$(x^2 + 3x - 4)(2x^2 - 5x + 3) = 0 \quad \text{or} \quad 3x^2 - 2x - 1 = 0.$$

The first equation is reduced to a solution of two quadratic equations

$$x^2 + 3x - 4 = 0 \quad \text{or} \quad 2x^2 - 5x + 3 = 0.$$

Using Viète's formulas, the solutions of the first equation are $x_1 = 1$, $x_2 = -4$; the solutions of the second equation are $x_3 = 1$, $x_4 = \frac{3}{2} = 1\frac{1}{2}$.

Finally, solving the third equation $3x^2 - 2x - 1 = 0$, gives $x_5 = 1$, $x_6 = -\frac{1}{3}$.

**Answer:** $1$, $1\frac{1}{2}$, $-4$, $-\frac{1}{3}$.

## Chapter 3

**Problem 16.** It is impossible that he was a Liar, because Liars always lie, so he would not tell the truth about himself. It is impossible that he was a Truth, because they never lie and by saying that he is a Liar, that person did lie. So, the only remaining option is he was one of the Slyboots (they lie sometimes and tell the truth sometimes). This time, he lied to the tourists. They will be better off finding another person to ask for directions. They can't completely trust this guy.

**Problem 17.** Assume the one who was talking to the tourists was a Truth. Then, it is impossible that he responded the way he did (he basically proclaimed that he is a Liar, which Truths just can't say). So, he has to be a Liar. Since he said that his friend is a Truth, his statement has to be a lie, and his friend has to be a Liar as well.

**Problem 18.** To solve this problem, we have to be very careful and pay attention to *all* the given information. We know that the tourists did not realize right away who was who after getting a response from the second person. The tourists were thinking for a few moments, trying to absorb the answer. So, the problem converts to a "perceptional" problem, similar to Problem 7 from Chapter 2. If the second person said "Yes" answering the tourists' question, they would not be able to determine who is who of the two friends. Indeed, you can get such a response from a Truth, but you can get it from a Liar as well. Therefore, the answer had to be "No". It implies that the second person is a Liar (if he would be a Truth, he would not say "No"). Therefore, the first person is a Truth.

**Problem 19.** Truths never lie, so it is impossible that the phone call was from a Truth; he would say that he lives in the Truths' district. The caller can't be a Slyboot, because they take turns telling truth and lies. If we assume that this was a Slyboot, we find that he was telling the truth twice about having a sick relative in his Slyboot family, and about the district where the family lives. Hence, the caller has to be a Liar, because he lied twice. He lied that the sick person is from his family, so the sick person has to be either a Truth or a Slyboot. The caller lied the second time as well when he said that the ambulance has to go to a Slyboot. It implies that the sick person has to be from the Truths' district and the ambulance has to go there.

**Problem 20.** The third person cannot be a Truth, because otherwise the second guy has to be a Truth, which is impossible (all three belong to different types; besides, he would have confirmed that fact instead of saying that he is a Slyboot). The second guy cannot be a Truth neither because he would have confirmed that he was a Truth, while answering the sergeant's question. So, the only option remaining for the first person is to belong to the Truths. Therefore, he was telling the truth saying that the second person is a Liar. Thus, the third person, clearly, is a Slyboot.

**Problem 21.** This problem has several alternative solutions.

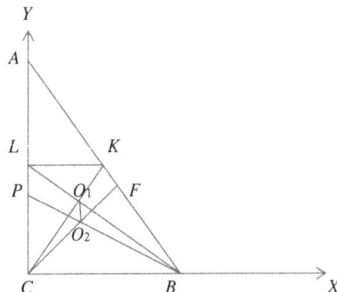

For example, when solving many geometrical problems that require finding angle measures or segment lengths, one usually looks for some triangles that have those elements, hoping to establish relationships that link triangles together, and lead to designing a plan for a solution. We leave this to the readers to explore whether to go in that or some other direction and find your own solution. We will take a different path here.

It is given that in the right triangle $ABC$ ($\angle C = 90°$), $AC = 12$, $BC = 9$, centroid $O_1$ is the point of intersection of the medians $BL$ and $CK$, and incenter $O_2$ is the point of intersection of the angle bisectors $CF$ and $BP$ of the angles $ACB$ and $CBA$, respectively. We have to find $O_1O_2$.

We will position the given triangle $ABC$ on the Cartesian coordinate plane in such a way that $C$ coincides with the origin, $A$ is located on the $Y$-axes, and $B$ is located on the $X$-axes. Now, we can define all the vertices by their coordinates: $C(0,0)$, $A(0,12)$, and $B(9, 0)$. $L$ is the midpoint of $AC$, so $L$ has the coordinates 0 and 6, $L(0,6)$. $LK$ is the midline in the triangle $ABC$, hence $LK \parallel BC$ and $LK = \frac{1}{2}BC = 4.5$. It implies that the coordinates of $K$ are 4.5 and 6, $K(4.5,6)$.

If we determine the coordinates of $O_1$ and $O_2$, we would be able to apply a general formula for finding the distance between two points given their coordinates,

$$O_1O_2 = \sqrt{(x_{O_2} - x_{O_1})^2 + (y_{O_2} - y_{O_1})^2},$$

and the problem will be solved.

Let's start with point $O_1$. By the properties of the centroid of a triangle, it divides each median in the ratio 2:3 from the vertices, so $CO_1 = 2O_1K$. We know the coordinates of $C$ and $K$ and we know the ratio in which $O_1$ divides $CK$. These facts should be sufficient enough to find the coordinates of $O_1$. To do so, let's solve a supplemental problem for establishing the coordinates of a point dividing the given segment in a specific ratio $t$.

Consider a general case of the three points $C(x_C, y_C)$, $K(x_K, y_K)$, and $O_1(x, y)$ such that $\overrightarrow{CO_1} = t \cdot \overrightarrow{O_1K}$. Then, the coordinates of the vectors are $\overrightarrow{CO_1} = (x - x_C, y - y_C)$, $\overrightarrow{O_1K} = (x_K - x, y_K - y)$, and, respectively,

$$x - x_C = t \cdot (x_K - x),$$

$$y - y_C = t \cdot (y_K - y).$$

Therefore, $x = \frac{t \cdot x_K + x_C}{t+1}$ and $y = \frac{t \cdot y_K + y_C}{t+1}$. So, substituting $t = 2$ and the coordinates of the points $C$ and $K$ gives $x = \frac{2 \cdot 4.5 + 0}{2+1} = 3$ and $y = \frac{2 \cdot 6 + 0}{2+1} = 4$. We get $O_1(3, 4)$. The next step is to find the coordinates of $O_2$.

$O_2$ is the point of intersection of the angle bisectors and it is equidistant from all the sides of the triangle. So, in fact, our goal is reduced to finding the radius of the inscribed circle in $ABC$. We will bring in one more supplemental problem and will prove that for any triangle, the radius of an inscribed circle equals the ratio of the triangle's area to its semi-perimeter, $r = \frac{S}{p}$, where $r$ is the inradius, $S$ is the area of $ABC$, and $p$ is the semi-perimeter of $ABC$.

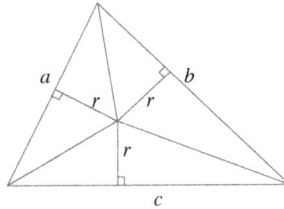

Connecting the vertices of a triangle to the center of the inscribed circle, we dissect it into three triangles with sides $a$, $b$, and $c$. The altitudes dropped to those sides from the common vertex at the center of the inscribed circle are all equal to the radius $r$. The area of each of the three triangles can be found as half the product of the base by the altitude dropped to that base:

$$S_1 = \frac{1}{2}ar,$$

$$S_2 = \frac{1}{2}br,$$

$$S_3 = \frac{1}{2}cr.$$

Adding the three equalities, we get the area of the original triangle

$$S = S_1 + S_2 + S_3 = \frac{1}{2}ar + \frac{1}{2}br + \frac{1}{2}cr = \frac{1}{2}r(a+b+c) = rp,$$

where $p = \frac{1}{2}(a+b+c)$ is the semi-perimeter of a triangle. So, we see that, indeed, $S = rp$, from which $r = \frac{S}{p}$.

The area of the right triangle $A\overset{\vee}{B}C$ can be found as half the product of its legs,

$$S = \frac{1}{2} \cdot 12 \cdot 9 = 54.$$

To find the semi-perimeter of $ABC$, we first apply the Pythagorean Theorem and find its hypotenuse $AB$, $AB = \sqrt{9^2 + 12^2} = 15$. Hence, $p = \frac{1}{2}(9 + 12 + 15) = 18$. Therefore, $r = \frac{S}{p} = \frac{54}{18} = 3$. Hence, the coordinates of $O_2$ are 3 and 3, $O_2(3,3)$. Indeed, dropping the perpendiculars from $O_2$ to $Y$ and $X$-axes, we'll find the distance from $O_2$ to both lines, which would equal $r$, the radius of the inscribed circle, which we just found as equal to 3. Finally, we got to the stage in our solution where we are ready to determine the distance between $O_1(3, 4)$ and $O_2(3,3)$. Finding the coordinates of $O_1$ and $O_2$, we see that they have the common abscissa 3, which proves that both points are located on the perpendicular to $X$-axes and the distance between them is easily found as the difference of their ordinates; there is no need to apply the general formula for the distance between two points, as we originally planned. Therefore, $O_1O_2 = 4 - 3 = 1$.

**Answer:** $O_1O_2 = 1$.

A few takeaways from this problem:

The decision to pick a path for the solution was derived from the question asked. Using Cartesian coordinates and vector algebra technique allowed us to design a plan that led to getting the desired distance between two points as a direct calculation based on their coordinates, skipping possible complicated searches for various relationships between the triangle's elements involving segment $O_1O_2$.

In our problem-solving process, we introduced two supplemental problems that clarified the next steps we needed to take and simplified the calculations: establishing formulas for the coordinates of a point dividing a segment in the given ratio and the expression for an inradius through the area and semi-perimeter of a triangle. Both give good examples of how instrumental supplemental problems are in overcoming intermediate obstacles in a problem-solving process.

An interesting outcome was obtained in passing as we determined the coordinates of $O_1$ and $O_2$ — because both points have the same abscissa, the segment $O_1O_2$ is perpendicular to side $BC$. This allowed us to simplify the final step in calculating the length of $O_1O_2$ as the difference of their ordinates; in fact, it equals the difference between the distances from each point to $BC$.

Finally, it merits pointing out that another reason we selected the Cartesian coordinates plane method here is because of its non-conventional nature. This powerful and useful technique is not well covered in high school curriculum. However, it's not difficult to use, and it may become very efficient not only in simplifying solutions, but in paving the way directly to answering the question asked.

## Chapter 4

**Problem 5.** We had on the left-hand side of the given equality $225 : 25 + 75$. By factoring out 25 properly, we should get $25 \cdot (\frac{9}{25} + 3)$, not $25 \cdot (9 : 1 + 3)$, as it was incorrectly applied in the proof. Since $25 \cdot (\frac{9}{25} + 3) = 84$, there is no contradiction with the calculation of the numerical value on the left side to be equal to the value on the right side.

**Problem 6.** Making our modifications, when we got to the point

$$4 \cdot (y + z - x) = 5 \cdot (y + z - x),$$

we then divided both sides by $(y + z - x)$, which is 0 according to the set values of $x$, $y$, and $z$. This is one of the commonly committed mistakes leading to a wrong conclusion. Canceling out a common factor on both sides of an equality, we can do it only when it's not 0.

**Problem 7.** It is important to understand that from $x = y$, it follows that $x^2 = y^2$, while the converse is not always correct. The true statement is, if $x^2 = y^2$, then $|x| = |y|$. Therefore, it is correct that $(2 - \frac{5}{2})^2 = (3 - \frac{5}{2})^2$, but it is wrong to deduce from this that $2 - \frac{5}{2} = 3 - \frac{5}{2}$. The correct outcome has to be $|2 - \frac{5}{2}| = |3 - \frac{5}{2}|$.

**Problem 8.** We can't apply to infinite series the same laws and operations applicable for finite sequences of numbers. This does not make any sense. The fallacy is that infinite series that are not convergent do not have a sum.

**Problem 9.** A similar reasoning as in Problem 8 — the length of a broken line is some finite number. Talking about the limit of that length brought us into infinite categories and comparing the finite amounts with infinite lengths leads to absurdity. The length of this broken line is a fixed number, and the discussed limit when the number of small triangles approaches infinity will not be equal to the hypotenuse. If at each vertex of the broken line we draw lines parallel to the legs, then it is obvious that the length of the broken line is equal to the sum of the legs regardless of how many vertices we have.

**Problem 10.** The point of intersection of angle bisector $BO$ and perpendicular bisector of $AC$ will never be inside of $ABC$.

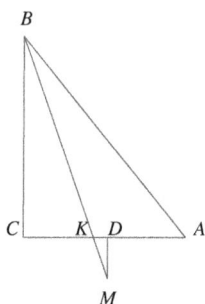

Let $K$ be the point of intersection of an angle bisector of angle $B$ and $AC$. By a well-known property of an angle bisector in a triangle, $K$ splits $AC$ in the ratio equal to the ratio of two other sides: $\frac{BC}{BA} = \frac{CK}{AK}$. In any right triangle, a hypotenuse is greater than the leg; so, $BC < BA$; therefore, $CK < KA$. It implies that $BK$ and the perpendicular bisector of $AC$ will intersect outside of the triangle $ABC$, so all the triangles we discussed simply do not exist.

**Problem 11.** When we raised both sides of the equality $\cos^2 \alpha = 1 - \sin^2 \alpha$ to the power $\frac{3}{2}$, then the obtained equality

$(\cos^2 \alpha)^{\frac{3}{2}} = (1 - \sin^2 \alpha)^{\frac{3}{2}}$ will be equivalent to the original only for such values of $\alpha$ for which $\cos \alpha \geq 0$. When $\alpha = \pi$, $\cos \alpha = -1$, therefore all the next manipulations have been done with the equality not equivalent to the original and it was impossible to get to the same numerical values after substituting $\pi$ for $\alpha$.

**Problem 12.** When we divided both sides of the inequality $2 \log \frac{1}{2} > 3 \log \frac{1}{2}$ by $\log \frac{1}{2}$, we had to reverse the sign of the inequality because $\log \frac{1}{2}$ is a negative number. So, the proper inequality after canceling out $\log \frac{1}{2}$ is $2 < 3$.

**Problem 13.** There was nothing wrong with the solution. The trick is in the statement of the problem, and not the solution. Before solving this problem, one has to find the range of permissible values of $x$:

$$\begin{cases} 45 - x \geq 0, \\ 9 - x \geq 0. \end{cases}$$

Solving the system of inequalities, we see that $x$ is restricted to values such that $x \leq 9$. Therefore, $\sqrt{45 - x} + \sqrt{9 - x} \geq \sqrt{36} + 0 = 6 > 4$. It is impossible that $\sqrt{45 - x} + \sqrt{9 - x} = 4$, as it was given in the problem. Before starting to solve a problem, make sure it is stated correctly.

**Problem 14.** One always has to be extra cautious while facing equations with parameters. The problem does not say that the equation has no other roots except $b$ and $c$. Therefore, we need to explore whether we indeed found *all* the possible values of $b$ and $c$, and did not miss any solutions. In fact, there exists one more root: $b = c = -\frac{1}{2}$.

Generally speaking, this quadratic equation will have the roots when its discriminant is a non-negative number, $D = b^2 - 4c \geq 0$ and $x = \frac{-b \pm \sqrt{b^2 - 4c}}{2}$. Knowing that $b$ and $c$ satisfy the equation, we can consider and analyze several scenarios when $b = \frac{-b \pm \sqrt{b^2 - 4c}}{2}$ or $c = \frac{-b \pm \sqrt{b^2 - 4c}}{2}$.

From the first equation, consider the case when $b = \frac{-b - \sqrt{b^2 - 4c}}{2}$. It follows that $3b = -\sqrt{b^2 - 4c}$. Squaring both sides and observing that since the right side is a negative number (square root of a number $n$ by definition is always a non-negative number, $\sqrt{n} \geq 0$) the left side has to be negative as well (so $b \leq 0$), we obtain that

$$9b^2 = b^2 - 4c, \text{ from which } c = -2b^2. \tag{1}$$

Similarly, solving the equation $c = \frac{-b+\sqrt{b^2-4c}}{2}$, we simplify it to $2c + b = \sqrt{b^2 - 4c}$, and after squaring both sides, we get $4c^2 + 4cb + b^2 = b^2 - 4c$. The last equation is further simplified to $c^2 + cb + c = 0$ or equivalently, $c(c + b + 1) = 0$. It follows that either $c = 0$ (the result we already derived before) or $c = -b - 1$. Substituting $c$ from equality (1) leads to $-2b^2 = -b - 1$. Now, solve the equation $2b^2 - b - 1 = 0$. It has two roots $b = 1$ and $b = -\frac{1}{2}$. As we observed above, only $b \leq 0$ satisfies our analysis, so we reject the positive root and get that $b = -\frac{1}{2}$. It follows that $c = -b - 1 = -(-\frac{1}{2}) - 1 = -\frac{1}{2}$. We invite the readers to investigate all other choices independently and justify that there are no other missing solutions.

**Problem 15.** The trick is that the rectangle's diagonal obscures what is going on. The pieces don't actually fit together in the way shown on the rectangle — there is a little bit of a gap (see the picture below). This gap is a parallelogram. Its area equals one square unit. The one square unit is "hidden" as it is depicted in the picture. If you cut the original square into the four pieces and then try to rearrange them to form a rectangle, the sides of the triangles and trapezoids will not fit together perfectly.

The dimensions of the rectangle are 5 and 13; the square's side has length 8. These are the consecutive numbers from the Fibonacci sequence. 5 is the fifth Fibonacci number, 8 is the sixth, and 13 is the seventh.

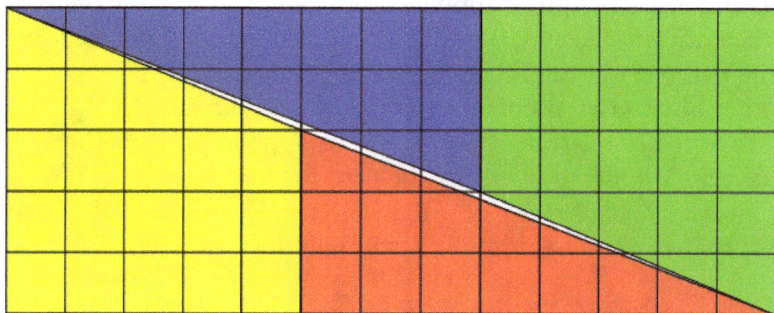

Note that the difference between the areas is $5 \cdot 13 - 8 \cdot 8 = 1$, or denoting the $n$th Fibonacci number as $F_n$, we can state that $F_5 \cdot F_7 - F_6^2 = 1$.

And, this is not a coincidence.

This paradox has an explanation based on the properties of the Fibonacci numbers. Let's first recall the definition of the Fibonacci sequence.

The Fibonacci sequence is an infinite sequence consisting of natural numbers such that $F_1 = 1, F_2 = 1$, and $F_{n+1} = F_n + F_{n-1}$ :

$$1,\ 1,\ 2,\ 3,\ 5,\ 8,\ 13,\ 21,\ 34,\ \dots\ .$$

Let's see if the similar relationship to $F_5 \cdot F_7 - F_6^2 = 1$ holds for the other Fibonacci numbers:

$$F_3 \cdot F_1 - F_2^2 = 2 \cdot 1 - 1 = 1,$$
$$F_4 \cdot F_2 - F_3^2 = 3 \cdot 1 - 4 = -1,$$
$$F_5 \cdot F_3 - F_4^2 = 5 \cdot 2 - 9 = 1,$$
$$F_6 \cdot F_4 - F_5^2 = 8 \cdot 3 - 25 = -1.$$

For the considered above examples, we see the pattern that $F_{n-1} \cdot F_{n+1} - F_n^2 = (-1)^n$. It holds for any three consecutive Fibonacci numbers and it is known as *Cassini's Fibonacci identity* after the Italian astronomer and mathematician Giovanni Domenico Cassini (1625–1712) who found it in 1680.

There are several proofs of this interesting property. To be consistent with our advice seeking distinct approaches to problems, we will show here two different proofs, one using the mathematical induction, and the other proof applying *Binet's Formula*, after the French mathematician Jacques Binet (1786–1856), for the explicit formula of the $n$th term in the Fibonacci sequence. We leave the exploration of other alternative proofs for the readers to pursue.

We will demonstrate the first proof applying the mathematical induction. We already verified the case for $n = 1$. Assume now that the identity holds for $n = k$, $F_{k-1} \cdot F_{k+1} - F_k^2 = (-1)^k$. We have to prove that in that assumption it will hold for $n = k + 1$, i.e., that

$$F_k \cdot F_{k+2} - F_{k+1}^2 = (-1)^{k+1}.$$

Recalling that by the definition of the Fibonacci numbers, $F_{k+2} = F_{k+1} + F_k$, we have

$$F_k \cdot F_{k+2} - F_{k+1}^2 = F_k \cdot (F_{k+1} + F_k) - F_{k+1}^2$$

$$= F_k \cdot F_{k+1} + F_k^2 - F_{k+1}^2 = F_{k+1}(F_k - F_{k+1}) + F_k^2$$

$$= F_{k+1} \cdot (-F_{k-1}) + F_k^2 = -F_{k-1} \cdot F_{k+1} + F_k^2$$

$$= -(F_{k-1} \cdot F_{k+1} - F_k^2) = -(-1)^k = (-1)^{k+1},$$

which was required to be proved.

In the second proof, we will be using *Binet's Formula* expressing the relationship between the Fibonacci numbers and the golden ratio, $\varphi = \frac{1+\sqrt{5}}{2}$:

$$F_n = \frac{\left(\frac{1+\sqrt{5}}{2}\right)^n - \left(\frac{1-\sqrt{5}}{2}\right)^n}{\sqrt{5}}.$$

First, note that $\left(\frac{1+\sqrt{5}}{2}\right) \cdot \left(\frac{1-\sqrt{5}}{2}\right) = \frac{1-5}{4} = -1$, so we can state that $\frac{1-\sqrt{5}}{2} = -\frac{1}{\varphi}$ and *Binet's Formula* can be rewritten as

$$F_n = \frac{\varphi^n - \left(-\frac{1}{\varphi}\right)^n}{\sqrt{5}}.$$

Therefore,

$$F_{n-1} \cdot F_{n+1} - F_n^2 = \frac{\varphi^{n-1} - \left(-\frac{1}{\varphi}\right)^{n-1}}{\sqrt{5}} \cdot \frac{\varphi^{n+1} - \left(-\frac{1}{\varphi}\right)^{n+1}}{\sqrt{5}} - \frac{\left(\varphi^n - \left(-\frac{1}{\varphi}\right)^n\right)^2}{(\sqrt{5})^2}$$

$$= \frac{\varphi^{2n-2} - (-1)^{n-1}}{\sqrt{5}\varphi^{n-1}} \cdot \frac{\varphi^{2n+2} - (-1)^{n+1}}{\sqrt{5}\varphi^{n+1}} - \frac{\varphi^{2n} - 2 \cdot (-1)^n + \left(-\frac{1}{\varphi}\right)^{2n}}{5}$$

$$= \frac{\varphi^{2n-2} + (-1)^n}{\sqrt{5}\varphi^{n-1}} \cdot \frac{\varphi^{2n+2} + (-1)^{n+2}}{\sqrt{5}\varphi^{n+1}} - \frac{\varphi^{4n} - 2 \cdot \varphi^{2n}(-1)^n + 1}{5\varphi^{2n}}$$

$$= \frac{\varphi^{4n} + \varphi^{2n-2} \cdot (-1)^{n+2} + (-1)^n \cdot \varphi^{2n+2} + (-1)^{2n+2} - \varphi^{4n} + 2 \cdot \varphi^{2n} \cdot (-1)^n - 1}{5\varphi^{2n}}$$

$$= \frac{\varphi^{2n-2} \cdot (-1)^{n+2} + (-1)^n \cdot \varphi^{2n+2} + 2 \cdot \varphi^{2n} \cdot (-1)^n}{5\varphi^{2n}}$$

$$= \frac{\varphi^{2n} \cdot (-1)^n ((-1)^2 \cdot \varphi^{-2} + \varphi^2 + 2)}{5\varphi^{2n}} = \frac{\varphi^{2n}(-1)^n(\varphi^{-2} + 2 + \varphi^2)}{5\varphi^{2n}}$$

$$= \frac{(-1)^n(\varphi + \frac{1}{\varphi})^2}{5} \tag{1}$$

Let's now simplify $\left(\varphi + \frac{1}{\varphi}\right)^2$ and then substitute the obtained result into (1).

$$\left(\varphi + \frac{1}{\varphi}\right)^2 = \left(\frac{\varphi^2 + 1}{\varphi}\right)^2 = \left(\frac{\left(\frac{1+\sqrt{5}}{2}\right)^2 + 1}{\frac{1+\sqrt{5}}{2}}\right)^2 = \left(\frac{\frac{6+2\sqrt{5}}{4} + 1}{\frac{1+\sqrt{5}}{2}}\right)^2$$

$$= \left(\frac{5 + \sqrt{5}}{1 + \sqrt{5}}\right)^2 = \left(\frac{\sqrt{5}(\sqrt{5} + 1)}{1 + \sqrt{5}}\right)^2 = (\sqrt{5})^2 = 5.$$

Substituting 5 into (1) gives $F_{n-1} \cdot F_{n+1} - F_n^2 = \frac{(-1)^n \cdot 5}{5} = (-1)^n$, which is the desired result. The similar paradox, therefore, can be created by using any triplet of consecutive Fibonacci numbers.

Recalling that for any Fibonacci numbers, $F_{n+1} = F_n + F_{n-1}$, we can rewrite this as $F_{n-1} = F_{n+1} - F_n$, and substituting this value of $F_{n-1}$ into *Cassini's Fibonacci identity*, we will get that $(F_{n+1} - F_n) \cdot F_{n+1} - F_n^2 = (-1)^n$, or equivalently, $F_{n+1}^2 - F_{n+1} \cdot F_n - F_n^2 = (-1)^n$. It's proven (try to prove by yourself!) that the equation $x^2 - x \cdot y - y^2 = \pm 1$ has no other roots in natural numbers. It is interesting that this was one of the facts (with some other properties of the Fibonacci numbers) allowing Soviet mathematician Yuri Matiyasevich to solve in 1970 the famous David Hilbert's Problem 10. It raised the question regarding the solvability of Diophantine equations (equations with two or more variables whose values are restricted to integers). Yuri Matiyasevich managed to prove that there is no uniform algorithm or technique to figuring out the solutions of such equations or even to determine if those solutions exist at all.

**Problem 16.** No segment vanished. Each of the 13 segments on the original figure was replaced by the respective 12 segments that are each one-twelfth longer making an impression that one segment disappears. The vanished line "diffused" among the remaining 12 segments increasing the length of each by $\frac{1}{12}$ of its length. The geometrical explanation relies on the fact that the cutting transversal line $AM$ across the parallel segments forms congruent alternate interior angles and, therefore, all the formed triangles are similar.

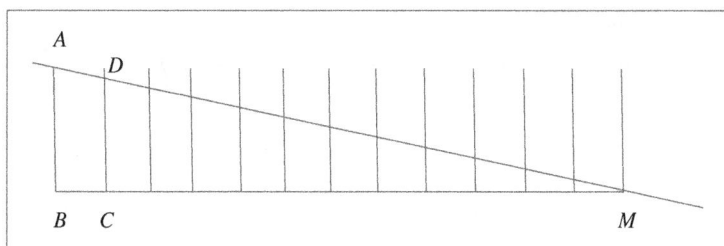

Without loss of generality, let's consider two triangles $ABM$ and $DCM$. Since $AB \parallel DC$ then $\angle A = \angle D$ and $\angle B = \angle C$. Angle $M$ is common in both triangles. We see that the triangles $ABM$ and $DCM$ are similar. It follows that $\frac{AB}{DC} = \frac{BM}{CM} = \frac{12}{11}$ (by construction, $BC = \frac{1}{12}BM$). So, $DC = \frac{11}{12}AB$. In other words, $AM$ cuts from the second segment next to $AB$ $\frac{1}{12}$ of its length. Similar relationships can be easily proved for the rest of the triangles in our configuration; $AM$ cuts from the third segment next to $AB$ $\frac{2}{12}$ of its length, from the fourth segment $\frac{3}{12}$ of its length, and so forth, up to the last segment from which it cuts $\frac{11}{12}$ of its length. When we slide the top piece to left one line, we put the cut piece of each segment, starting from the second, to the preceding segment. And, since each cut piece is greater by $\frac{1}{12}$ than the preceding piece, each segment increases by $\frac{1}{12}$ of its length. This small increase is not easily caught by the eye, creating the mysterious "disappearance" of one segment out of 13 on the picture.

# Chapter 5

**Problem 7.** The first East's move must be in hearts. Since West has no hearts at all, he would hit it with spades-trumps, reducing his trumps to three, while South has four spades-trumps on hand. No matter how West plays, he cannot have more than six tricks (four spades, the ace of clubs, and the ace of diamonds) if East and South continue making their moves in hearts. It worth noting that West will definitely win a 9-trick game having his hand first. He would collect three of spades making the first three moves with the ace, king, and queen of spades. The fourth move has to be either the ace of diamonds or the ace of clubs; it's irrelevant. If he goes with the ace of diamonds, for example, then the next move has to be the king of diamonds, which South has to hit with his last remaining trump. No matter what the next move South makes, the trick will be West's because he has one more trump on hand. So, all the next tricks will be his.

**Problem 8.** South has to go with the ace of diamonds. This will be his first trick. The second move must be in spades. This trick will be taken by East with his ace of spades. In the third move, East has to go with diamonds "killing" West's queen of diamonds with the South's remaining trump. In total, West will have only seven tricks.

**Problem 9.** Let's answer the second question first. If East would go first, he would win an eight-trick game by collecting five diamonds (he has to make the first two moves with the ace and king of diamonds), the ace and king of hearts, and the ace of spades. However, he would lose the game if West is on the eldest hand. West has to go twice with clubs: first with the ace of clubs collecting the clubs trick, and then he has to go with the clubs again, with the eight of clubs, for instance. To get the second trick, East has to hit the club with either the ace or king of diamonds. If he puts any other diamond card, South would collect the trick because he has the jack of diamonds. Since West has the queen and seven of diamonds, he will be able to collect a trick in diamonds for his queen, no matter how East would play next. Therefore, in total East would give in three tricks, that is, one of the diamonds, the seven of spades, and the nine of clubs. Even if East collects a trick with the ace of hearts and then goes into the king of hearts hitting into West's trumps, this would not change the ultimate game's outcome. In this case, East still will give in three tricks, the king of hearts, the seven of spades, and the nine of clubs. There is an important takeaway from this game. While announcing the number of tricks, one should not hope and rely on the desired possible favorable cards disposition. Should East have announced 7 tricks in diamonds, he would not have lost the game.

## Chapter 6

**Problem 12.**

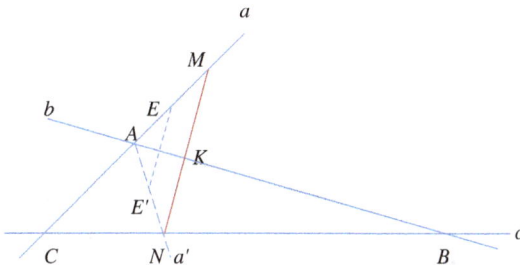

We are given three straight lines intersecting each other, $a \cap b = A$, $b \cap c = B$, and $a \cap c = C$. Our goal is to construct a segment with end points on $a$ and $c$ perpendicular to $b$ and such that its midpoint is located on $b$.

The problem will be solved if we construct the image of $a$ in reflection through $b$ axis till its intersection at some point with $c$ and then find the image of that point on $a$ in the same reflection through $b$ axis. Connecting these two points will give us the desired segment.

**Construction steps:** Select a random point $E$ on $a$ and find its image $E'$ in reflection through $b$ axis. Draw $AE'$ (this is the straight line $a'$, the image of $a$ in reflection through $b$ axis) till intersection with $c$ at $N$. Find $M$, the image of $N$ in reflection through $b$ axis. $M$ must be located on $a$, the reverse image of $a'$ in reflection through $b$ axis, because every point on $a'$ is reflected into the respective point located on $a$. $MN$ intersects $b$ at $K$, which has to be the midpoint of the segment $MN$ due to reflection properties, and by construction, $MN \perp b$. Therefore, $MN$ is the desired segment.

This problem has the unique solution because there is only one point of intersection of the image of $a$ in reflection through $b$ axis with $c$. Respectively, there is only one segment perpendicular to $b$ drawn from $N$ till its intersection with $a$ at $M$, its image in reflection through $b$ axis. By definition of the line symmetry, the midpoint $K$ of $MN$ must be located on the line of symmetry $b$.

**Problem 13.**

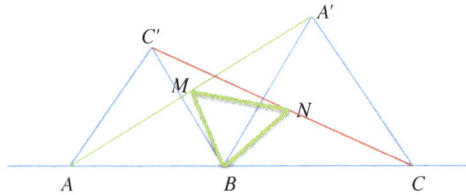

Consider a 60° clockwise rotation about the center $B$. Since both triangles $AC'B$ and $BA'C$ are equilateral triangles, this rotation will take $A$ into $C'$ and $A'$ into $C$. So, $AA'$ is rotated into $C'C$. It implies that the midpoint of $AA'$ will be rotated into the midpoint of $CC'$, that is, $M$ into $N$. Therefore, since $M$ and $N$ are the images of each other in a 60° rotation about the center $B$, $BM = BN$ and $\angle MBN = 60°$. It follows that $\triangle MBN$ is the equilateral triangle, which was to be proved.

**Problem 14.**

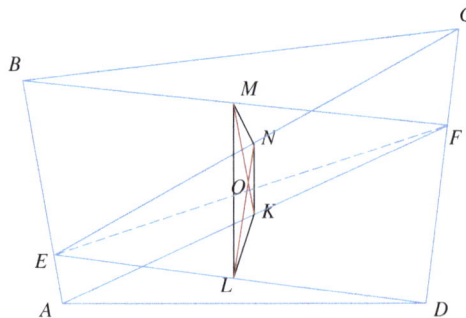

Let $M$ be the midpoint of $BF$, $N$ the midpoint of $CE$, $K$ the midpoint of $AF$, and $L$ the midpoint of $DE$. Observe that $MK$ is the midline of $\triangle AFB$, so $MK \parallel BA$ and $MK = \frac{1}{2}AB$; $NL$ is the midline of $\triangle CED$, so $NL \parallel CD$ and $NL = \frac{1}{2}CD$.

Recall that in a convex quadrilateral, all interior angles are less than 180° and the two diagonals both lie inside the quadrilateral.

To prove that $MNKL$ is a convex quadrilateral, it suffices to prove that its diagonals $MK$ and $NL$ intersect.

Consider a homothety with the center $E$ and coefficient $k = 2$. It will take $NL$ into $CD$. Draw $EF$ and let $O$ be its point of intersection with $NL$. It follows that $F$ is the image of $O$ in our homothety with center $E$ and coefficient $k = 2$. Then, $EO = OF$, i.e., $O$ is the midpoint of $EF$.

Consider now the homothety with the center $F$ and coefficient $k = 2$. It will take $MK$ into $BA$. Assume that $EF$ intersects $MK$ at point $O'$. $E$ has to be the image of $O'$ in this homothety, and $EO' = O'F$, i.e., $O'$ is the midpoint of $EF$. But, a segment can have only one midpoint; therefore, $O$ and $O'$ coincide, which implies that $MK$ and $NL$ indeed intersect, and $O$ is the point of their intersection. It follows that $MNKL$ is a convex quadrilateral. We have to prove now the second statement that the area of $MNKL$ does not depend on the selection of $E$ and $F$ on the sides of $ABCD$.

The area of a convex quadrilateral can be calculated as one-half of the product of its diagonals by sine of the angle between them, that is,

$$S_{MNKL} = \frac{1}{2} MK \cdot NL \cdot \sin \alpha,$$

where $\alpha$ is the angle between $MK$ and $NL$. As we observed at the beginning of our proof, $MK \parallel AB$ and $MK = \frac{1}{2} AB$, and $NL \parallel CD$ and $NL = \frac{1}{2} CD$. The angles between pairs of respective parallel lines are congruent, so we see that $\alpha$ is the same angle as the angle between $AB$ and $CD$. Thus, it does not depend on the location of $E$ on $AB$ and $F$ on $CD$. So, the area of $MNKL$ is calculated as $S_{MNKL} = \frac{1}{2} \cdot \frac{1}{2} AB \cdot \frac{1}{2} CD \cdot \sin \alpha = \frac{1}{8} AB \cdot CD \cdot \sin \alpha$, and indeed, it is the same constant number regardless of the selection of $E$ on $AB$ and $F$ on $CD$, dependent only on lengths of sides $AB$ and $CD$ and the angle between them in the original convex quadrilateral.

**Problem 15.**

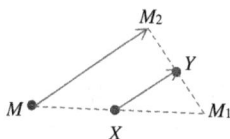

Generally speaking, two consecutive point symmetries with centers $X$ and $Y$ produce the same outcome as translation by vector $2\overrightarrow{XY}$, that is, $M_2$, the image of $M$ in two consecutive point symmetries with the centers $X$ and $Y (M \rightarrow M_1 \rightarrow M_2)$, is the same point $M_2$, the image of $M$, in translation by vector $2\overrightarrow{XY}$. Indeed, by the definition of point symmetry, $MX = XM_1$ and $M_1Y = YM_2$, so $XY$ is the midline in the triangle $MM_1M_2$. It implies that $XY = \frac{1}{2}MM_2$ and $XY \parallel MM_2$, which proves our statement.

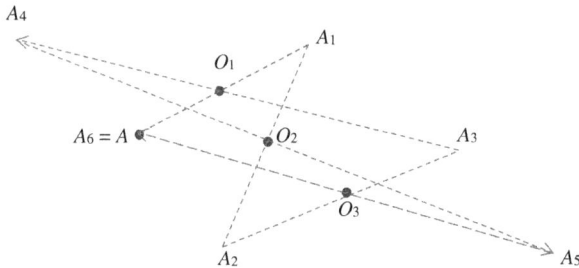

Keeping in mind the above observation, let's now list the translation equivalent outcomes of our point symmetries transformations:

$A$ into $A_2$ — translation by vector $2\overrightarrow{O_1O_2}$;
$A_2$ into $A_4$ — translation by vector $2\overrightarrow{O_3O_1}$;
$A_4$ into $A_6$ — translation by vector $2\overrightarrow{O_2O_3}$;

where we denote by $A_6$ the final image in all our point symmetries transformations (it is the image of $A_5$ in the last point symmetry with the center $O_3$). Our goal is to prove that $A$ and $A_6$ coincide.

In the last step, we considered $O_2O_3$, which is the midline in the triangle $A_6A_4A_5$. Therefore,

$$O_2O_3 \parallel A_6A_4 \quad \text{and} \quad O_2O_3 = \frac{1}{2}A_6A_4. \tag{1}$$

$O_1$ divides $AA_1$ and $A_4A_3$ in half; therefore, $AA_4A_1A_3$ is a parallelogram. So, $A_1A_3 \parallel AA_4$ and $A_1A_3 = AA_4$.

$O_2O_3$ is the midline in the triangle $A_1A_2A_3$. Therefore, $O_2O_3 \parallel A_1A_3$ and

$$O_2O_3 = \frac{1}{2}A_1A_3 = \frac{1}{2}AA_4. \tag{2}$$

Comparing (1) and (2), we see that we have two segments $A_6A_4$ and $AA_4$ of the same length drawn from $A_4$ in the same direction and parallel to $O_2O_3$. It is possible only when $A$ and $A_6$ coincide.

**Problem 16.**

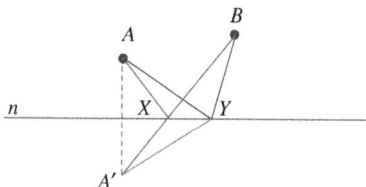

Assume $n$ is the given straight line, and $A$ and $B$ are the given points outside of $n$. We have to determine (construct) the point on $n$ such that the sum of distances from it to $A$ and $B$ is the minimal from any other point located on $n$.

Construct $A'$ as the image of $A$ in reflection through $n$ axis and connect it with $B$. Let $X$ be the point of intersection of $A'B$ and $n$. $X$ is the desired point. To prove it, we pick any other random point $Y$ on $n$, and will prove that $AX + XB < AY + YB$.

By the properties of reflection, $AX = A'X$. Therefore, $AX + XB = A'X + XB = A'B$. The triangle inequality theorem states that for any triangle, the sum of the lengths of any two sides must be greater than the length of the remaining side. So, in $\triangle A'BY$, the sum of two sides $A'Y$ and $YB$ is greater than the third side $A'B$, $A'Y + YB > A'B$.

Observing that $A'Y = AY$ due to the properties of reflection, we get the desired result,

$$AX + XB = A'B < AY + YB.$$

**Problem 17.** Let's analyze the problem and see what strategy we can apply for our construction. In order to construct a triangle, we require the measure of three parts of the triangle. Given the lengths of the three medians, most likely, we would need to establish some triangle with three sides constructible under the given conditions and then see how to proceed from there to our goal.

So, assume we have the desired triangle $ABC$ with the three given medians $AA_1, BB_1$, and $CC_1$. Denote the point of their intersection, the centroid, by $O$.

Consider translation by vector $\overrightarrow{OA_1} = \frac{1}{3}\overrightarrow{AA_1}$. Let $K$ be the image of $A_1$ in such a translation. Then, $OA_1 = A_1K$, and, as given, we know that $A_1$ is the midpoint of $BC$.

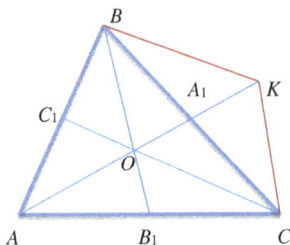

Therefore, $A_1$ divides the diagonals $OK$ and $BC$ of $OBKC$ in half. It follows that $OBKC$ is a parallelogram. So, $KC \parallel BO$, $OC \parallel BK$ and $KC = BO$, $OC = BK$. As the result of our translation, we obtained triangle $OBK$ with all three sides of the known lengths. Indeed, recalling the property of the centroid to divide the medians in the ratio of two-thirds from any vertex to the midpoint of the opposite side, we have $BK = OC = \frac{2}{3}CC_1$, $BO = \frac{2}{3}BB_1$, and $OK = \frac{2}{3}AA_1$ (since $OK = 2OA_1 = 2 \cdot \frac{1}{3}AA_1$). Hence, triangle $OBK$ is constructible by the three sides, each of which is two-thirds of the given respective medians. As we construct the triangle $OBK$, it will not be hard to complete our construction and get to the triangle $ABC$. We will need to locate $A_1$ the midpoint of $OK$ and do two point reflections with the centers $A_1$ and $O$ transforming $B$ into $C$ and $K$ into $A$, respectively. We will arrive at the desired triangle $ABC$. Clearly, this problem will have the solution when the given lengths of the medians satisfy the triangle inequality, i.e., the sum of any two of them is greater than the third median.

We leave the actual constructions to be done by the readers.

**Problem 18.**

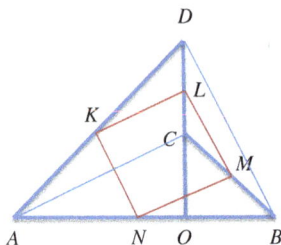

We are given two right isosceles triangles $AOD$ ($AO = DO$) and $BOC$ ($BO = CO$). Points $K$, $L$, $M$, and $N$ are the midpoints of $AD$, $CD$, $CB$, and $AB$, respectively. We have to prove that $KLMN$ is a square. First, notice that $KN$ and $LM$ are the midlines in triangles $ABD$ and $CBD$. Hence, each of these segments is parallel to $BD$ and half as long as $BD$. Similarly, $KL$ and $MN$ are the midlines in triangles $ACD$ and $ACB$, respectively. So, $KL \parallel MN \parallel AC$ and $KL = MN = \frac{1}{2}AC$. A 90° clockwise rotation about $O$ takes $A$ into $D$ and $C$ into $B$, since we are given that $AOD$ and $BOC$ are the right isosceles triangles. Thus, this rotation takes $AC$ into $DB$. By the properties of rotations, this implies that $AC = BD$ and the angle between them is 90°, i.e., $AC \perp BD$. It follows that $KL$, $MN$, $KN$, and $ML$ have equal lengths as halves of the equal segments ($AC = BD$) and $KL \perp LM$ and $KN \perp NM$. Therefore, $KLMN$ is a square, as was required to be proved.

## Chapter 8

## Problem 11.

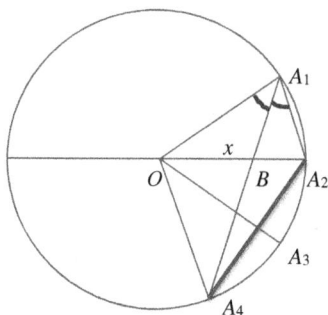

We proved that $\angle OA_1A_2 = 72°$. It is given that $A_1A_4$ is the angle bisector of the angle $OA_1A_2$. So, $\angle OA_1A_4 = \angle A_4A_1A_2 = 72° : 2 = 36°$. The inscribed angle $\angle A_4A_1A_2$ is subtended by the chord $A_4A_2$. Since the inscribed angle is equal to one half of the central angle subtended by the chord, we have that angle $\angle A_4OA_2$ is equal to twice the angle $\angle A_4A_1A_2$, $\angle A_4OA_2 = 36° \cdot 2 = 72°$. This implies that $A_4A_2$ is the side of the regular pentagon. Let's find its length.

Applying the Law of Cosines to the triangle $A_4OA_2$ gives
$A_4A_2{}^2 = A_4O^2 + A_2O^2 - 2\cdot1\cdot1\cos\angle A_4OA_2 = 2 - 2\cos72° = 2(1 - \cos72°)$ $(1)$.
We need to modify and evaluate now the expression $(1 - \cos72°)$:

$$1 - \cos72° = \underbrace{\sin^2 36° + \cos^2 36°}_{=1} - \underbrace{(\cos^2 36° - \sin^2 36°)}_{=\cos72°}$$

$$= 2\sin^2 36° = 2(1 - \cos^2 36°).$$

To find $\cos36°$, we consider triangle $A_1OA_2$ in which $\angle A_1OA_2 = 36°$, $OA_1 = OA_2 = 1$, and $A_1A_2 = \frac{-1+\sqrt5}{2}$ (this value was determined while solving Problem 10 in Chapter 8). Applying the Law of Cosines to the triangle $A_1OA_2$ and expressing $\cos36°$, we get

$$\cos36° = \frac{A_1O^2 + A_2O^2 - A_1A_2{}^2}{2\cdot A_1O\cdot A_2O} = \frac{1 + 1 - \left(\frac{-1+\sqrt5}{2}\right)^2}{2}$$

$$= \frac{2 - \frac{3-\sqrt5}{2}}{2} = \frac{1+\sqrt5}{4}.$$

Therefore, $\cos36° = \frac{1+\sqrt5}{4}$. Respectively,
$$1 - \cos72° = 2(1 - \cos^2 36°) = 2\left(1 - \left(\frac{1+\sqrt5}{4}\right)^2\right) = 2\left(1 - \frac{3+\sqrt5}{8}\right) = \frac{5-\sqrt5}{4}$$
and substituting this into $(1)$ gives
$A_4A_2{}^2 = 2\,(1 - \cos72°) = 2\cdot\frac{5-\sqrt5}{4} = \frac{5-\sqrt5}{2}$, and we arrive at the desired length of the side of the regular pentagon inscribed into the unitary circle to be equal to $\sqrt{\frac{5-\sqrt5}{2}}$, $A_4A_2 = \sqrt{\frac{5-\sqrt5}{2}}$.

**Problem 12.**

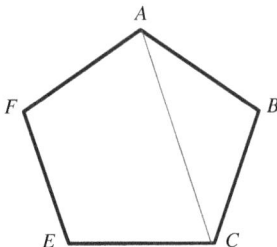

In a regular pentagon, the angle between sides equals $108°$. It is given that $AB = BC = 1$. Therefore, applying the Law of Cosines to triangle $ABC$, we have

$$AC^2 = AB^2 + BC^2 - 2 \cdot AB \cdot BC \cdot \cos 108° = 2 - 2\cos 108°$$
$$= 2(1 - \cos 108°) = 2(1 - \cos(180° - 72°)) = 2(1 + \cos 72°). \quad (1)$$

We found in the previous problem that $1 - \cos 72° = \frac{5-\sqrt{5}}{4}$, from which $\cos 72° = \frac{\sqrt{5}-1}{4}$. Substituting this value into (1) gives $AC^2 = 2\left(1 + \frac{\sqrt{5}-1}{4}\right) = \frac{3+\sqrt{5}}{2}$.

Hence,

$$AC = \sqrt{\frac{3+\sqrt{5}}{2}} = \sqrt{\frac{6+2\sqrt{5}}{4}} = \sqrt{\frac{1+2\sqrt{5}+\left(\sqrt{5}\right)^2}{4}}$$
$$= \sqrt{\frac{(1+\sqrt{5})^2}{4}} = \frac{1+\sqrt{5}}{2}.$$

As we mentioned before, $\frac{1+\sqrt{5}}{2}$ is the famous golden ratio. So, the diagonal's length of our regular pentagon equals the golden ratio!

**Problem 13.** Using the results obtained during the solution of Problem 10, it's easy to get that $r = a \cdot \frac{1+\sqrt{5}}{2}$, where $r$ is the radius of the circumcircle and $a$ is the side of a regular decagon. Indeed, referring to the picture for Problem 10, and applying the Law of Cosines to triangle $A_1OA_2$, we have $a^2 = r^2 + r^2 - 2r^2 \cdot \cos 36°$. Therefore,

$$r^2 = \frac{a^2}{2(1 - \cos 36°)} = \frac{a^2}{2\left(1 - \frac{1+\sqrt{5}}{4}\right)} = \frac{a^2}{\frac{3-\sqrt{5}}{2}}$$
$$= \frac{2a^2(3+\sqrt{5})}{(3-\sqrt{5})(3+\sqrt{5})} = \frac{2a^2(3+\sqrt{5})}{9-5} = \frac{a^2(6+2\sqrt{5})}{4}$$
$$= \frac{a^2(1+2\sqrt{5}+(\sqrt{5})^2)}{4} = a^2\left(\frac{1+\sqrt{5}}{2}\right)^2,$$

from which $r = a \cdot \frac{1+\sqrt{5}}{2}$.

Since it is given that $a = 1$, we obtain that $r = \frac{1+\sqrt{5}}{2}$.

**Problem 14.**

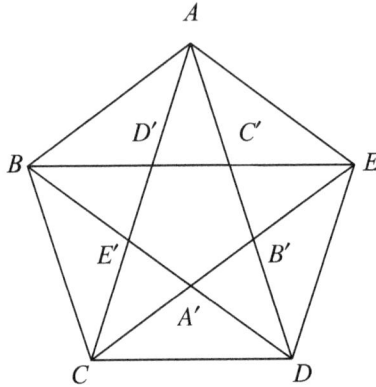

It is given that the length of a side of the regular pentagon $ABCDE$ equals 1. Our plan will be to find the length of the side of the regular pentagon $A'B'C'D'E'$ and then apply the theorem about the ratios of the areas of similar polygons: *The ratio of the areas of the two similar polygons is equal to the squared ratio of their corresponding linear elements.* So, first consider triangle $AC'E.\,AE = 1$, $\angle AC'E = 108°$, and $AC' = C'E$ (hopefully, readers can easily prove this). We already found that $\cos 72° = \frac{\sqrt{5}-1}{4}$; thus, we obtain that $\cos 108° = \cos(180° - 72°) = -\cos 72° = -\frac{\sqrt{5}-1}{4} = \frac{1-\sqrt{5}}{4}$.

Let $x = AC' = C'E$. Applying the Law of Cosines to the triangle $AC'E$ gives $AE^2 = x^2 + x^2 - 2x^2 \cos 108°$. Substituting $AE = 1$ and expressing $x$, we get $1 = 2x^2(1 - \cos 108°) = 2x^2(1 - \frac{1-\sqrt{5}}{4})$, so

$$x = \sqrt{\frac{1}{2(1 - \frac{1-\sqrt{5}}{4})}} = \sqrt{\frac{4}{6+2\sqrt{5}}} = \sqrt{\frac{4}{1+2\sqrt{5}+(\sqrt{5})^2}}$$

$$= \sqrt{\frac{4}{(1+\sqrt{5})^2}} = \frac{2}{1+\sqrt{5}}.$$

By checking angles, we see that $\triangle D'AC'$ is similar to $\triangle CAD$, so $\frac{D'C'}{CD} = \frac{AC'}{AD}$. Our goal is to find $D'C'$. We can get it from the last equality, substituting the values of $CD = 1$, $AD = \frac{1+\sqrt{5}}{2}$ (as it was calculated in Problem 12), and $AC' = \frac{2}{1+\sqrt{5}}$. We have

$$D'C' = \frac{CD \cdot AC'}{AD} = \frac{1 \cdot \frac{2}{1+\sqrt{5}}}{\frac{1+\sqrt{5}}{2}}$$

$$= \frac{4}{(1+\sqrt{5})^2} = \frac{4}{6+2\sqrt{5}} = \frac{2}{3+\sqrt{5}}.$$

Finally, the ratio of the areas of similar regular pentagons has to equal the square of the ratio of their respective sides. Hence, denoting the area of $ABCDE$ by $S$ and the area of $A'B'C'D'E'$ by $S'$, we obtain that

$$\frac{S}{S'} = \left(\frac{1}{\frac{2}{3+\sqrt{5}}}\right)^2 = \frac{(3+\sqrt{5})^2}{4} = \frac{14+6\sqrt{5}}{4} = \frac{7+3\sqrt{5}}{2}.$$

**Problem 15.**

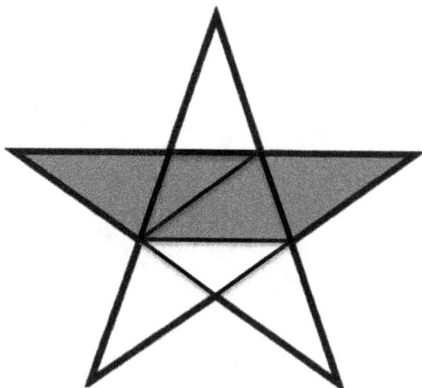

As you can see in the picture above, the painted and unpainted portions of the star can be cut into a set of the congruent triangles.

Therefore, the areas of these portions will be the same, and each equals half of the area of the star.

**Problem 16.**

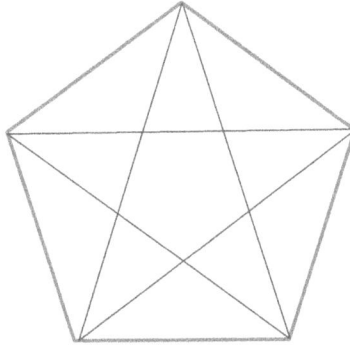

While working with previous problems, we noticed a few important relations between elements in the regular pentagon. We will exploit them here. We cut $EBN'AP$ from our star and moved it into $BFM'GN'$ placing $EB$ onto $BF$. The new polygon is centrosymmetric with its center of symmetry point $O$, the intersection of $BC$ and $EF$.

Let's prove it.

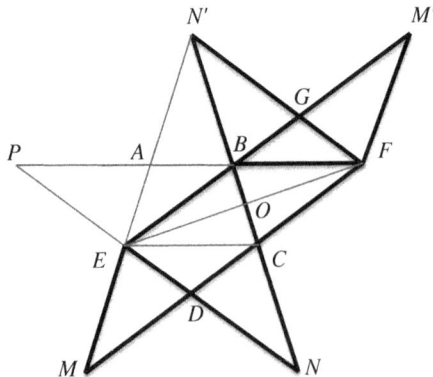

Considering angles formed by the diagonals of the regular pentagon, it is easy to get that $AB \parallel EC$ and $BE \parallel CD$. It implies that $BF \parallel EC$ and $BE \parallel CF$. Therefore, $EBFC$ is a parallelogram. Furthermore, since $BE = EC$, we conclude that $EBFC$ is a rhombus. Its point of symmetry is $O$, the point of intersection of its diagonals $EF$ and $BC$. We also see that in pentagons $EBN'AP$, BFM'GN', and $ECNDM$, all the respective angles and sides are equal. It follows that by cutting $EBN'AP$ from our star and placing $BE$ on top of $BF$, $EBN'AP$ will coincide with $BFM'GN'$, which is congruent to the bottom pentagon $ECNDM$. We obtained the new polygon $EMDNCFM'GN'B$. This concave polygon consists of the rhombus and two congruent pentagons attached to its opposite sides, so it is centrosymmetric with the point of symmetry $O$.

**Problem 17.** One example of such a quadrilateral can be the quadrilateral obtained by connecting any four vertices of the regular pentagon, for instance, the quadrilateral $ABCD$, as shown in the figure below.

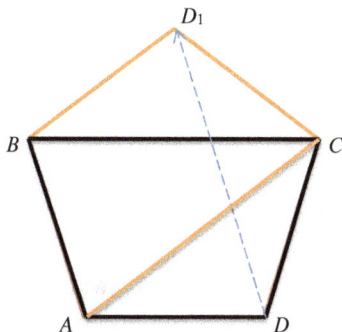

By moving $D$ into $D_1$ we get the new quadrilateral $ABD_1C$ congruent to $BADC$. One can get several such quadrilaterals by moving $D$ or any other vertex of $ABCD$ into the other vertex of the regular pentagon on the picture.

We suggest that readers consider the detail rigorous proof of this interesting problem.

# Chapter 9

## Problem 7.

| | | | | | | | |
|---|---|---|---|---|---|---|---|
| 54 | 49 | 40 | 35 | 56 | 47 | 42 | 33 |
| 39 | 36 | 55 | 48 | 41 | 34 | 59 | 46 |
| 50 | 53 | 38 | 57 | 62 | 45 | 32 | 43 |
| 37 | 12 | 29 | 52 | 31 | 58 | 19 | 60 |
| 28 | 51 | 26 | 63 | 20 | 61 | 44 | 5 |
| 11 | 64 | 13 | 30 | 25 | 6 | 21 | 18 |
| 14 | 27 | 2 | 9 | 16 | 23 | 4 | 7 |
| 1 | 10 | 15 | 24 | 3 | 8 | 17 | 22 |

The above solution was suggested by the prominent Swiss mathematician Leonhard Euler (1707–1783) in 1757. He was one of the first to study this problem. In his letter to German mathematician Christian Goldbach (1690–1764), Euler wrote about finding the solution to the problem, but did not explain the details.

The knight's tour problem is an example of the more general *Hamiltonian path problem* in *graph theory*. On an $8 \times 8$ board, there are 26,534,728,821,064 directed closed tours (two tours along the same path that travel in opposite directions are counted separately). Hamilton path problems are beyond the scope of our book, but ambitious readers may find lots of interesting facts in other literature. One such work, for example, is Gordon Horsington's book *Century/Acorn User Book of Computer Puzzles*, Dally Simon ed (1984), describing a computer program that finds a knight's tour for any starting position using Warnsdorff's rule.

## Chapter 10

**Problem 11.** In a trapezoid $ABCD (AB \parallel CD), AB = a$, and $CD = b$. $MN$ is parallel to the bases of the trapezoid and divides it into two trapezoids of equal areas. Find the length of $MN$.

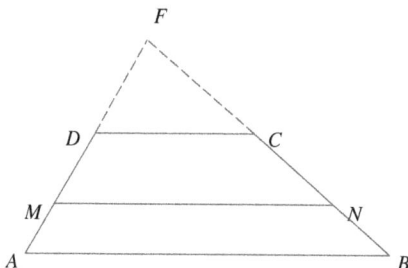

This interesting problem about expressing the length of $MN$ as the quadratic mean of the trapezoid's bases was examined in Chapter 6 of *Geometrical Kaleidoscope*, Dover Publications, 2017. Chapter 6 was devoted to the applications of the theorem of the ratios of the areas of similar polygons. We extended the sides $AD$ and $BC$ till their intersection at $F$, forming similar triangles $DFC$, $MFN$, and $AFB$. The solution was based on considering the ratios of areas of several pairs of similar triangles and expressing the requested linear element, segment $MN$, in terms of the corresponding given linear elements $a$ and $b$. We believe it should be a good exercise for the readers to recreate this proof on their own (or you can refer to the solution mentioned above in *Geometrical Kaleidoscope*).

Let's go over another two solutions to this problem.

### Solution 1.

In this solution, we will use the formula for the area of a trapezoid as half the product of the sum of its bases by its altitude, and apply it to setting up the equation based on the equality of the areas of trapezoids $AMNB$ and $MDCN$. This is a straightforward approach. There will be no special tricks or auxiliary elements introduced.

First, let's draw $DE \perp AB$. Denote $F$ the intersection of $DE$ and $MN$. By construction, $DE$ is the altitude of the trapezoid $ADCB$, $DF$ is the altitude of the trapezoid $MDCN$, and $FE$ is the altitude of the trapezoid $AMNB$. For convenience, we introduce several variables:

$$MN = x, \ DF = h_1, \ FE = h_2, \ DE = h.$$

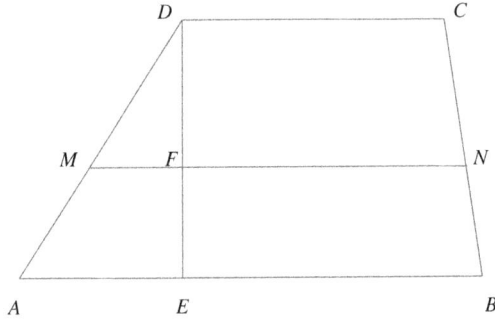

The area of $ADCB$ equals $S_{ADCB} = h \cdot \frac{a+b}{2}$ .

The area of $MDCN$ equals $S_{MDCN} = h_1 \cdot \frac{b+x}{2}$.

The area of $AMNB$ equals $S_{AMNB} = h_2 \cdot \frac{a+x}{2}$.

It is given that $S_{MDCN} = S_{AMNB} = \frac{1}{2}S_{ADCB}$. Therefore, $h_1 \cdot \frac{b+x}{2} = h_2 \cdot \frac{a+x}{2}$, from which it follows that

$$\frac{h_1}{h_2} = \frac{a+x}{b+x}. \tag{1}$$

The area of $ADCB$ equals the sum of the areas of $MDCN$ and $AMNB$. Observing also that by construction, $h = h_1 + h_2$, we can express the given information as

$$(h_1 + h_2) \cdot \frac{a+b}{2} = h_1 \cdot \frac{b+x}{2} + h_2 \cdot \frac{a+x}{2}.$$

Simplifying the above equation, we get

$$ah_1 + bh_1 + ah_2 + bh_2 = bh_1 + xh_1 + ah_2 + xh_2.$$

Canceling the like terms gives $ah_1 + bh_2 = xh_1 + xh_2$, from which $ah_1 - xh_1 = xh_2 - bh_2$, or equivalently, $h_1(a - x) = h_2(x - b)$, and finally,

$$\frac{h_1}{h_2} = \frac{x - b}{a - x}.$$

Substituting in the last equation the expression for $\frac{h_1}{h_2}$ from equality (1) gives

$$\frac{a + x}{b + x} = \frac{x - b}{a - x}.$$

It follows that $a^2 - x^2 = x^2 - b^2$, from which we get $x^2 = \frac{a^2 + b^2}{2}$ and, respectively, $x = \sqrt{\frac{a^2 + b^2}{2}}$.

The problem is solved.

## Solution 2.

Here, we will be looking for similar triangles to utilize the equality of the ratios of the respective elements for setting up the equation for $x$. It is important to emphasize that the trapezoids $AMNB$ and $MDCN$ into which $MN$ splits $ADCB$ are not similar trapezoids. Therefore, we will be making use of some additional constructions allowing us to identify and select similar figures. Dealing with the trapezoid, the quadrilateral with two parallel sides, and the given segment $MN$ parallel to its bases, it's natural to draw one more parallel line, $CK \parallel DA$, and apply the properties of the obtained several parallelograms.

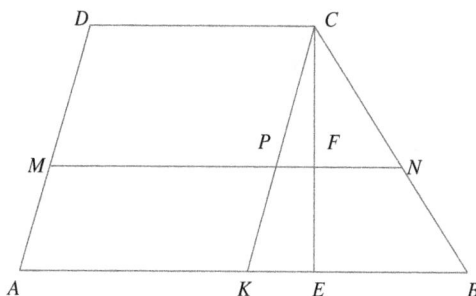

Next, we draw $CE \perp AB$. Denote by $P$ and $F$, respectively, the points of intersection of $CK$ with $MN$ and $CE$ with $MN$. By drawing $CK \parallel DA$ we get parallelograms $ADCK$, $MDCP$, and $MPKA$, with all pairs of the respective opposite sides being equal:

$$MP = AK = DC = b \text{ and, respectively,}$$

$$PN = MN - MP = x - b \text{ and } KB = AB - AK = a - b.$$

$CE \perp AB$ and $MN \parallel AB$; therefore, $CF \perp PN$. So, $CF$ is the altitude in triangle $PCN$, $CE$ is the altitude in triangle $KCB$.

Let's denote $CF = h_1$, $FE = h_2$, and $CE = h$. Observing that triangles $KCB$ and $PCN$ are similar (I hope it can be easily proved by the readers), we get that $\frac{CF}{CE} = \frac{PN}{KB}$, or equivalently, $\frac{h_1}{h_1+h_2} = \frac{x-b}{a-b}$.

Dividing the numerator and denominator on the left side by $h_1$ gives $\frac{1}{1+\frac{h_2}{h_1}} = \frac{x-b}{a-b}$. Recalling from Solution 1 equality (1), $\frac{h_1}{h_2} = \frac{a+x}{b+x}$, we will rewrite it as $\frac{h_2}{h_1} = \frac{b+x}{a+x}$. Now, substituting the expression for $\frac{h_2}{h_1}$ into the last equation, we get

$$\frac{1}{1 + \frac{b+x}{a+x}} = \frac{x-b}{a-b}.$$

Making simplifications leads to

$$\frac{a+x}{a+b+2x} = \frac{x-b}{a-b},$$

$$(a+x)(a-b) = (a+b+2x)(x-b),$$

$$a^2 + xa - ab - xb = ax + bx + 2x^2 - ab - b^2 - 2xb.$$

Canceling the like terms, we get $a^2 + b^2 = 2x^2$, from which we obtain the desired result as $x = \sqrt{\frac{a^2+b^2}{2}}$.

**Problem 12.**
**Algebraic solution.**

Denote by $x$ one of the sides of the rectangle, and by $p$ its semi-perimeter. Then, the second side of the rectangle can be expressed as $(p - x)$. For the area of the rectangle, we have $S = x(p - x)$. Modifying the last expression, we get the quadratic equation for $x$

$$x^2 - px + S = 0.$$

Solving this equation gives $x = \frac{p}{2} \pm \sqrt{\frac{p^2}{4} - S}$. Clearly, the last equality holds only for $\frac{p^2}{4} - S \geq 0$ or, equivalently, when $S \leq \frac{p^2}{4}$. The maximum value of $S$ is attained when $S_{\text{max}} = \frac{p^2}{4}$. The respective value of $x$ then will be $x = \frac{p}{2} \pm \sqrt{\frac{p^2}{4} - S} = \frac{p}{2}$. Therefore, this rectangle becomes a square and indeed it has the greatest area of all rectangles with the same perimeter, which is what had to be proved.

**Geometrical solution.**

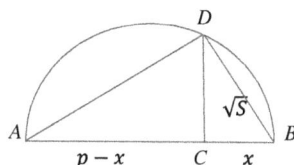

Let's draw a semicircle with the diameter $AB = p$, where $p$ is the semi-perimeter of the rectangle. Cut $AB$ into two segments $AC$ and $CB$ such that $AC = p - x$ and $CB = x$. Draw $DC \perp AB$ where $D$ belongs to the semicircle. Then, by the properties of inscribed angles subtended by the diameter, $\angle ADB = 90°$ (see Chapter 9). $DC$ is the altitude dropped from the vertex of the right angle to the hypotenuse. Hence, $DC = \sqrt{AC \cdot CB}$ (this property is easy to prove from the similarity of the right triangles in which $\triangle ADB$ is cut by its altitude $DC$). So, we see that $DC = \sqrt{(p - x) \cdot x}$. On the other hand, we know that the area of the rectangle is $S = x(p - x)$. It follows that $DC = \sqrt{S}$. It is obvious that the maximum value of $DC$ is attained when it equals the radius of the semicircle. Then, the maximum value of $S$ is attained when $S_{\text{max}} = \frac{p^2}{4}$, which is possible for $p - x = x$ or equivalently, for $x = \frac{p}{2}$. The last equality implies that the rectangle becomes a square. We obtained that a square has the greatest area of all rectangles with the same perimeter, as it was requested to be proved.

## Chapter 11

**Problem 1.** Divide nine coins in three sets of three coins each. Select any pairs of three coins and place them on a balance weight scale. This is your first weighing. If the pans balance, then the counterfeit is in the remaining set of three coins not selected for weighing. Select any two of those three coins and weigh them. If the pans balance, the remaining coin is counterfeit; if, on the other hand, a pan goes up, the lighter coin is counterfeit.

If in our first weighing the pans are not in balance and one pan goes up, we select the lighter set of three coins and repeat the step two.

The case of 27 coins is solved in a similar fashion. Divide 27 coins in three sets of nine coins each. Select any pairs of nine coins and place them on a balance weight scale. This is your first weighing. Identify the lighter set containing a counterfeit. After first weighing, you will be left with nine coins, one of which is a counterfeit. You will have two more weighings to find it. Follow all the steps indicated in the first part of the problem.

**Problem 2.** It is easy to check that knowing the sum of any two weights we can identify the weight of each in any combination of the pairs of the given weights.

Consider all the pairs of weights with 1,000 grams being one of them. We had the following summary weights: 2,001 grams (1000 + 1001), 2,002 grams (1000 + 1002), 2,004 grams (1000 + 1004), and 2,007 grams (1000 + 1007). Assume we select any pair of weights and weigh it. If the pair's weight will not be one of the listed above, we know for sure that there is no 1,000-gram weight in the selected pair. We then select any other pair from the remaining three weights and check its weight. In two weighings of randomly selected pairs, we can identify the pair with 1,000 grams in it (or if none, then the remaining weight will be the desired weight of 1,000 grams). In the last remaining third weighing of one of the weights in the identified pair, we can determine 1,000-gram weight.

**Problem 3.** The robbers have to identify the bag with counterfeits. The idea of how to do it with just one weighing is not obvious at all. However, it is clear that first they need to come up with some method for distinguishing bags from each other. The key idea is in assigning each bag a number. Use a pen and number each bag from 1 to 9. Then, get from each bag the number of coins equal to this bag's number, that is, one coin from the first bag, two coins from the second bag, and so on up to the ninth bag, from which they need to take nine coins. They get total of $1+2+3+4+5+6+7+8+9 = 45$ coins. The weight of genuine 45 coins is supposed to be $45 \cdot 10 = 450$ grams. After weighing of 45 selected coins, they will get the smaller weight because there are counterfeits among 45 coins, and they are lighter. The difference between 450 grams and the result of weighing will indicate the number of the bag with counterfeit coins.

Indeed, assume, for example, that bag #3 has all counterfeits. Each of the three coins, taken from that bag, is 1 gram lighter than the genuine coin. Then, the weight of 45 coins will be 447 grams. The difference $450 - 447 = 3$ gives the number of counterfeits on the scale, which in turn indicates bag #3 as the bag with counterfeits.

**Problem 4.** Let $a$ be the length of the left weighing scale shoulder and $b$ be the length of the right shoulder. According to Archimedes' Law of the Lever,

> *Magnitudes are in equilibrium at distances reciprocally proportional to their weights.*

## LEVER RULE

- *With Fulcrum at P, weights $W_A$ and $W_B$ at the end of a lever, for equilibrium, the lever rule states:*

$$W_A / W_B = b/a$$

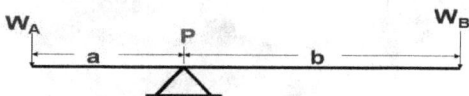

The weight was supposed to be 1 pound in each case. The store owner placed a 1-pound weight on the left weighing pan in the first weighing, and he placed a 1-pound weight on the right pan for the second customer's order. Having $W_a = 1$, from the equality $\frac{W_a}{W_b} = \frac{b}{a}$, it follows that $W_b = \frac{a}{b}$. In other words, when there is a 1-pound weight on the left pan, there is a weight of $W_b = \frac{a}{b}$ pounds of the straw-berries placed on the right pan (under these conditions, the scale will be in balance). A similar outcome is attained in the second case, when there is a 1-pound weight on the right pan. There is a weight of $W_a = \frac{b}{a}$ pounds on the other pan to get the scale in balance. So, the customer got in total $\frac{a}{b} + \frac{b}{a}$ pounds of strawberries. Let's now evaluate this value and find the difference $\frac{a}{b} + \frac{b}{a} - 2 = \frac{a^2+b^2-2ab}{ba} = \frac{(a-b)^2}{ba}$. Clearly, $\frac{(a-b)^2}{ba} > 0$, because the nominator is a positive number as a square of some number, and the denominator is a positive number as product of two positive numbers.

It implies that $\frac{a}{b} + \frac{b}{a} > 2$. Therefore, the customer got more than 2 pounds of strawberries. The store owner is hurting his business by using this weighing scale.

**Problem 5.** Place any two coins on the right pan and one coin along with a 5-gram scale weight on the left pan. If the pans balance, the counterfeit is the remaining coin. Take all the genuine coins out from pans and place the counterfeit on one pan while keeping a 5-gram scale weight on the other pan to see if counterfeit's weight is less or greater than 5 grams.

If in our first weighing the weight of two coins is greater than the weight of another coin and the 5-gram scale weight, we then need to place each of them on the two pans to compare their weights. If they have equal weights, then the coin located in the first weighing along with the 5-gram scale weight on another pan is counterfeit.

And, we found out as well that it weighs less than the genuine coin. If on the contrary, one of the coins in our second weighing had greater weight than the other, then it has to be the counterfeit and its weight is greater than 5 grams. In any case, just in two weighings, we succeeded to identify the counterfeit and determine if its weight is more or less than 5 grams.

Now, assume that in our first weighing the weight of two coins is less than the weight of another coin and the 5-gram weight. In the next weighing, we need to place the two coins from the first pan on different pans and see which one has greater weight. If they have the same weight, then the coin located on the pan with the 5-gram weight in the first weighing is counterfeit and its weight is greater than 5 grams. If on the contrary, in the second weighing one of the coins has less weight, then it is counterfeit and its weight is less than 5 grams. We accounted now for all possible scenarios and explained how to find the counterfeit and determine if it is heavier or lighter than a genuine coin.

**Problem 6.** Place 50 coins on each pan. If the pans are in balance, the remaining coin is a counterfeit. In the second weighing, place a genuine coin on the left pan and the counterfeit on the right pan and see which one is heavier.

If in the first weighing the pans are not in balance, then divide 50 coins from the lighter in weight pile into two piles of 25 coins in each and place them on the pans. If the pans are in balance, then the counterfeit is in the first pile of 50 coins and it is heavier than a genuine coin. If the pans are not in balance, then the counterfeit is one of these coins and it is lighter than a genuine coin.

**Problem 7.**

a) $a = 6$, $b = 9$, $c = 1$.

$$
\begin{array}{r}
6 \\
+\ 99 \\
6 \\
\hline
111
\end{array}
$$

b)    $97$
   $\times\ 11$
   $\overline{\phantom{0}97}$
   $\underline{97\phantom{00}}$
   $\overline{1067}$

c)    $115$
   $\times\ \ 98$
   $\overline{\phantom{0}920}$
   $\underline{1035\phantom{0}}$
   $\overline{11270}$

d) In Russian: $\partial = 4$, $в = 5$, $a = 9$, $ч = 2$, $e = 1$, $m = 0$, $ы = 6$, $p = 8$.

   In Polish: $a = 4$, $c = 7$, $d = 8$, $e = 3$, $r = 1$, $t = 9$, $w = 5$, $y = 6$, $z = 2$.

   In Spanish: $a = 8$, $c = 3$, $d = 5$, $o = 6$, $r = 9$, $s = 4$, $t = 0$, $u = 1$.

**Problem 8.** Let's add and subtract the same product of $1987 \cdot 19871987$ and regroup the addends in the following way:

$$1987 \cdot 19861986 - 1986 \cdot 19871987 = 1987 \cdot 19861986$$

$$- 1987 \cdot 19871987 + 1987 \cdot 19871987 - 1986 \cdot 19871987$$

$$= 1987 \cdot (19861986 - 19871987) + 19871987 \cdot (1987 - 1986)$$

$$= 1987 \cdot (-10001) + 19871987 \cdot 1 = 19871987 - 1987 \cdot (10000 + 1)$$

$$= 19871987 - 19870000 - 1987 = 19871987 - 19871987 = 0.$$

**Problem 9.** We can replace each fraction in the given sum as the difference of two fractions $\frac{1}{n \cdot (n+1)} = \frac{1}{n} - \frac{1}{n+1}$ and get

$$\frac{1}{2} + \frac{1}{2 \cdot 3} + \frac{1}{3 \cdot 4} + \cdots + \frac{1}{99 \cdot 100} = \left(1 - \frac{1}{2}\right) + \left(\frac{1}{2} - \frac{1}{3}\right)$$

$$+ \left(\frac{1}{3} - \frac{1}{4}\right) + \cdots + \left(\frac{1}{99} - \frac{1}{100}\right) = 1 - \frac{1}{2} + \frac{1}{2} - \frac{1}{3} + \frac{1}{3} - \frac{1}{4}$$

$$+ \cdots + \frac{1}{99} - \frac{1}{100} = 1 - \frac{1}{100} = \frac{99}{100}.$$

**Problem 10.** To simplify this expression, we need to multiply the numerator and denominator of each fraction by the conjugate of the number in each denominator and use the formula for the difference of squares:

$$\frac{1}{\sqrt{1}+\sqrt{2}} + \frac{1}{\sqrt{2}+\sqrt{3}} + \frac{1}{\sqrt{3}+\sqrt{4}} + \cdots + \frac{1}{\sqrt{99}+\sqrt{100}}$$

$$= \frac{\sqrt{2}-\sqrt{1}}{(\sqrt{1}+\sqrt{2})(\sqrt{2}-\sqrt{1})} + \frac{\sqrt{3}-\sqrt{2}}{(\sqrt{2}+\sqrt{3})(\sqrt{3}-\sqrt{2})}$$

$$+ \cdots + \frac{\sqrt{100}-\sqrt{99}}{(\sqrt{99}+\sqrt{100})(\sqrt{100}-\sqrt{99})}$$

$$= \frac{\sqrt{2}-\sqrt{1}}{2-1} + \frac{\sqrt{3}-\sqrt{2}}{3-2} + \cdots + \frac{\sqrt{100}-\sqrt{99}}{100-99}$$

$$= \sqrt{2} - \sqrt{1} + \sqrt{3} - \sqrt{2} + \cdots + \sqrt{100} - \sqrt{99}$$

$$= \sqrt{100} - \sqrt{1} = 10 - 1 = 9.$$

**Problem 11.** Alternative solutions:

$$12 - 3 - 4 + 5 - 6 + 7 + 89 = 100,$$

$$12 + 3 + 4 + 5 - 6 - 7 + 89 = 100,$$

$$123 - 45 - 67 + 89 = 100,$$

$$123 + 4 - 5 + 67 - 89 = 100,$$

$$123 - 4 - 5 - 6 - 7 + 8 - 9 = 100.$$

**Problem 12.**

$$\frac{\text{XXII}}{\text{VII}} = \pi$$

Taking one of the matches in the nominator and placing it above the two matches on the right-hand side, we got the "equality" $22 : 7 = \pi$. Since $\pi \approx 3.1415$, we arrived at the desired result.

**Problem 13.** Yes, it is (7 in Roman numerals is written as VII):

$$\overline{\phantom{xx}}\!\!\!\times\!\!\text{II}\phantom{xx}$$

**Problem 14.** Let's assume that the four digits $a$, $b$, $c$, and $d$ are such that after adding them at the right to 9999, the new eight-digit number $\overline{9999abcd}$ becomes a square of some natural number $x$, that is, $x^2 = \overline{9999abcd}$.

Observe that $10,000^2 = 100,000,000$ is the smallest nine-digit natural number, which is a square of an integer, and $9999^2 = (10,000 - 1)^2 = 100,000,000 - 20,000 + 1 = 99,980,001$ is the greatest eight-digit number which is the square of an integer. In our assumption, $9999^2 < x^2 < 10000^2$, which implies $9999 < x < 10000$, since $x$ has to be a natural number. The last inequality is impossible for a natural $x$. Therefore, we arrive at the conclusion that it is impossible to find four digits $a$, $b$, $c$, and $d$ such that $x^2 = \overline{9999abcd}$.

**Problem 15.** The hint here is in the word ROME, pointing to Roman numerals:

$$
\begin{array}{r}
DCVI \\
- \quad LXV \\
\hline
DXLI
\end{array}
$$

In Arabic numerals, it is $606 - 65 = 541$.

**Problem 16.** The sum of the digits of the number $\underbrace{111\ldots1}_{81\,\text{digits}}$ is 81, so it is divisible by 9. Dividing 111111111 by 9, gives 12345679. Therefore, dividing $\underbrace{111\ldots1}_{81\,\text{digits}}$ by 9 gives, $\underbrace{12345679\ldots12345679}_{9\,\text{times}}$ i.e., the quotient is the number consisting of 12345679 repeated nine times. The sum of digits of this number is

$$
1 \cdot 9 + 9 \cdot (2 + 7) + 9 \cdot (3 + 6) + 9 \cdot (4 + 5) + 9 \cdot 9
$$
$$
= 9 \cdot (1 + 9 + 9 + 9 + 9) = 9 \cdot 37,
$$

which is obviously divisible by 9. Therefore, we arrive at the conclusion that $\underbrace{111...1}_{81\,\text{digits}}$ is divisible by $9^2 = 81$.

**Problem 17.** To solve the problem, we need to complete the square and then represent the given expression as the difference of squares, $a^2 - b^2 = (a - b)(a + b)$, which would enable us to factor it. In order to do that, let's add and subtract $2^{992}$:

$$2^{1982} + 1 = (2^{1982} + 2^{992} + 1) - 2^{992} = ((2^{991})^2 + 2 \cdot 2^{991} + 1) - 2^{992}$$
$$= (2^{991} + 1)^2 - (2^{496})^2 = (2^{991} + 1 - 2^{496})(2^{991} + 1 + 2^{496}).$$

**Problem 18.** The given number can be written as $10^{1995} + 1 = (10^{665})^3 + 1$. Now, we can factor the obtained sum as the sum of cubes by using the formula

$$a^3 + b^3 = (a + b)(a^2 - ab + b^2).$$

We have $(10^{665})^3 + 1 = (10^{665} + 1)((10^{665})^2 - 10^{665} + 1)$, which is a composite number since it has two factors distinct from the original number and 1.

**Problem 19.** Let $x$ be the number of oranges one can buy for \$1. Then, the cost of 25 oranges is $x$ dollars. According to the first of the given conditions, the cost of 1 orange is $\frac{1}{x}$ dollars, while according to the second condition, the same cost of 1 orange is $\frac{x}{25}$ dollars. We get the equation $\frac{1}{x} = \frac{x}{25}$. It simplifies to $x^2 = 25$, from which we select only the positive root of $x = 5$. So, one can buy five oranges for \$1, and then 15 oranges can be purchased for \$3.

**Problem 20.** Let $x$ be the boy's age now. Then, he was $(x - 2)$ years old 2 years ago, and he will be $(x + 2)$ years old in 2 years. According to the problem, $2(x - 2) = x + 2$. Solving this equation gives $x = 6$.

Analogously, letting the girl's age now be $y$, we get that she was $(y - 3)$ years old 3 years ago, and she will be $(y + 3)$ years old in 3 years. According to the problem, $3(y - 3) = y + 3$. Solving this equation gives $y = 6$.

As we can see now, both are of the same age.

**Answer:** The boy and the girl are 6 years old each.

**Problem 21.** Let $x$ be the digit in the tens place and $y$ be the digit in the unit place. Then, the number can be written as $10x + y$, and the sum of its digits is $x + y$. We need to solve the following equation:

$$10x + y = 2(x + y),$$

which simplifies to $8x = y$. Since $x$ and $y$ are digits, clearly, the only feasible value for $x$ is $x = 1$. It follows that $y = 8$. Therefore, this number is 18.

**Answer:** The number is 18.

**Problem 22.** The problem is about the speed covered by each hand and the distance between them (in minutes) after specific time passed. The distance (in minutes) each hand travels is measured by the minute markers of the clock. At 2 pm, the distance between the minute and hour hands is 10 minutes. Comparing the speeds of the minute and hour hands, we know that the minute hand moves 12 times as fast as the hour hand.

Assume the minute hand will overtake the hour hand in $x$ minutes after 2 pm.

Then, at the meeting point, the distance covered by the minute hand is $x$ minutes and the distance covered by the hour hand is $\frac{1}{12}x$ minutes. The difference of the distances covered by each hand is the original distance between them at 2 pm, i.e., 10 minutes. Hence, we get the equation $x - \frac{1}{12}x = 10$. It follows that $\frac{11}{12}x = 10$, from which $x = 10\frac{10}{11}$ minutes. Therefore, after 2 pm, the clock hands will overlap at $2:10\frac{10}{11}$ pm.

**Problem 23.** Let $x$ be the number of three-headed dragons, $y$ be the number of 40-legged dragons, and $n$ be the number of legs each three-headed dragon has. Since the creatures have 26 heads in total, we arrive at the first equation $3x + y = 26$. They have a total of 298 legs; therefore, we can set up the second equation $xn + 40y = 298$. So, our goal is to solve the system of linear equations

$$\begin{cases} 3x + y = 26, \\ xn + 40y = 298. \end{cases}$$

Multiplying the first equation by $-40$ and adding to the second equation gives

$$x \cdot (120 - n) = 742.$$

Since $x$ has to be a natural number, the only possible values it can attain are the factors of 742, that is, it can be 1, 2, 7, or 53 (because $742 = 1 \cdot 2 \cdot 7 \cdot 53$).

Let's examine each of those values.

Assuming that $x = 53$, we get a contradiction with the first equality $3x + y = 26$. Indeed, $3 \cdot 53 + y = 26$, which leads to $y = -133$. This is impossible because $y$ is a natural number.

Assuming $x = 1$ and substituting it in the equality $x \cdot (120 - n) = 742$, we have $1 \cdot (120 - n) = 742$, from which $n = -622$. This is impossible because $n$ is a natural number.

Assuming $x = 2$ and substituting it in the equality $x \cdot (120 - n) = 742$, we have $2 \cdot (120 - n) = 742$, from which $n = -251$. This is also impossible because $n$ is a natural number.

Verifying $x = 7$ (there are 7 dragons on the island), we see that it is the only valid option, and substituting it into $x \cdot (120 - n) = 742$, we get $7 \cdot (120 - n) = 742$, which yields $n = 14$.

**Answer:** Each three-headed dragon has 14 legs.

**Problem 24.** Let $x$ be the number of big bags and $y$ be the number of smaller size bags having 10 bags inside them. Then, the total number of bags is $10 + 10 \cdot (x + y)$. It is given that $x + y = 54$, so there are a total of $10 + 10 \cdot 54 = 550$ bags.

**Answer:** 550 bags.

**Problem 25.**

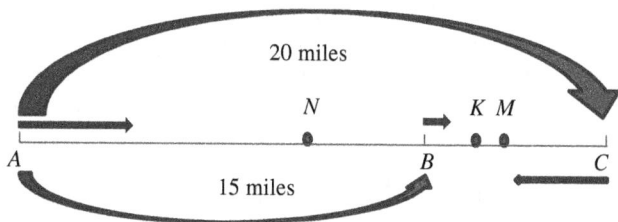

Let's start with drawing the diagram interpreting the problem's conditions. Car is driving in a direction from $A$ to $B$. Pedestrian is walking in a direction from $B$ to $C$. Bus is driving in a direction from $C$ to $A$ facing the car and the pedestrian. Assume the car overtook the pedestrian at point $M$, the car and the bus met at point $N$, and the bus and the pedestrian met at point $K$. Now, we introduce several variables: let $x$ miles/h be the car's speed, $y$ miles/h — the pedestrian's speed, $z$ miles/h — the bus's speed, and $t$ hours — the time when the car overtook the pedestrian.

The distance covered by the car till point $M$ was $xt$ miles, the distance covered by the pedestrian till point $M$ was $yt$ miles. The difference between these distances is 15 miles (in terms of the segments on our diagram, this distance can be expressed as $AM - BM = AB = 15$). So, we can set up the first equation

$$xt - yt = 15.$$

The car and the bus met half hour earlier than the car overtook the pedestrian, so the car and the bus spent $(t - \frac{1}{2})$ hours driving before they met at point $N$. The distance covered by the car till $N$ was $x \cdot (t - \frac{1}{2})$ miles. The distance covered by the bus till $N$ was $z \cdot (t - \frac{1}{2})$ miles. Since they were driving facing each other, the total distance covered by each till meeting at point $N$ was 20 miles, the distance between $A$ and $C$. This can be expressed by the equation

$$x \cdot \left(t - \frac{1}{2}\right) + z \cdot \left(t - \frac{1}{2}\right) = 20.$$

Finally, the pedestrian and the bus met in $\frac{1}{3} \cdot (t - \frac{1}{2})$ hours according to the problem's conditions. The distance covered by the pedestrian till meeting point $K$ was $\frac{1}{3} \cdot (t - \frac{1}{2})y$ miles. The distance covered by the bus till meeting point $K$ was $\frac{1}{3} \cdot (t - \frac{1}{2})z$ miles. They were also facing each other; therefore, the distance covered by both till point $K$ was 5 miles (it is $BK + KC = 5$). So, the third equation is

$$\frac{1}{3} \cdot \left(t - \frac{1}{2}\right) y + \frac{1}{3} \cdot \left(t - \frac{1}{2}\right) z = 5.$$

Our goal now is to solve the following system of equations:

$$\begin{cases} xt - yt = 15, \\ x \cdot \left(t - \frac{1}{2}\right) + z \cdot \left(t - \frac{1}{2}\right) = 20, \\ \frac{1}{3} \cdot \left(t - \frac{1}{2}\right) y + \frac{1}{3} \cdot \left(t - \frac{1}{2}\right) z = 5. \end{cases}$$

It can be rewritten as

$$\begin{cases} t(x - y) = 15, \\ (t - \frac{1}{2})(x + z) = 20, \\ (t - \frac{1}{2})(y + z) = 15. \end{cases}$$

To solve this system, we subtract the third equation from the second to get

$$\left(t - \frac{1}{2}\right)(x - y) = 5$$

and then divide it by the first equation to obtain

$$\frac{\left(t - \frac{1}{2}\right)(x - y)}{t(x - y)} = \frac{5}{15}.$$

Canceling out $(x - y)$ (clearly, $x \neq y$; so, we can do it), the last equation simplifies to $\frac{t - \frac{1}{2}}{t} = \frac{1}{3}$, solving which we get that $t = \frac{3}{4}$ hour or $t = 45$ minutes.

**Answer:** In 45 minutes, the car will overtake the pedestrian.

**Problem 26.** We can reasonably assume that her age is expressed as a two-digit number $10a + b$. By inserting 0 between the digits of a two-digit number, we will get a three-digit number expressed as $100a + b$. So, her age is $(10a + b)$ and her great-grandpa's age is $(100a + b)$. Therefore, we get the equation $6 \cdot (10a + b) = 100a + b$. It follows that $40a = 5b$, from which $b = 8a$. Since each of the digits $a$ and $b$ accepts the values from 0 to 9, the only feasible value of $a$ is 1, $a = 1$. Then, $b = 8$. Therefore, Liana is 18 years old.

**Problem 27.**

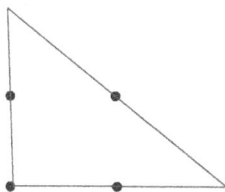

**Problem 28.**

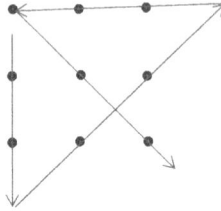

**Problem 29.** Alternative three solutions:

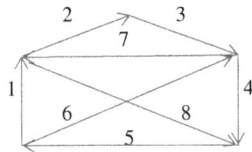

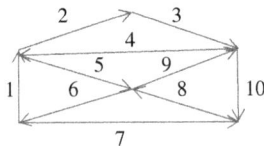

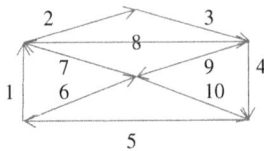

**Problem 30.**

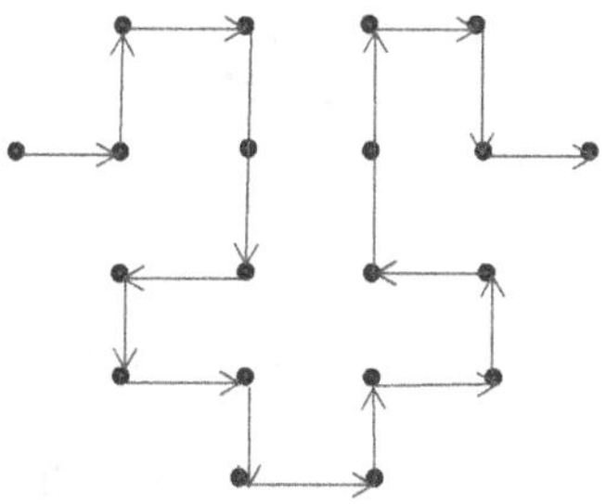

**Problem 31.** Counting vertices of a star and points of intersection of the segments connecting the vertices, we have a total of 10 points located on five segments with four points on each.

**Problem 32.**

**Problem 33.**

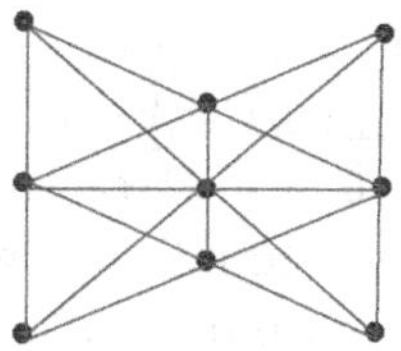

**Problem 34.**

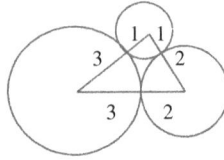

Radii equal 1, 2, and 3. Then, the sides of the triangle are 4, 3, and 5.

It is indeed a right triangle because $4^2 + 3^2 = 5^2$.

**Problem 35.** Yes, it is possible.

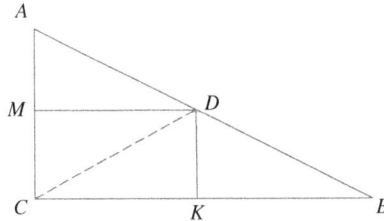

Consider the right triangle $\triangle ABC$ ($\angle C = 90°$).

Connect the vertex of the right angle $C$ with the midpoint $D$ of the hypotenuse $AB$. $D$ is the center of the circumcircle of the triangle $ABC$ (see Chapter 9); $CD = AD = DB$.

It follows that we obtained two isosceles triangles $ADC$ and $CDB$. It's not hard to prove that they have equal areas.

Let's draw the perpendiculars from $D$ to $AC$ and $BC$, $DM \perp AC$ and $DK \perp BC$. Then, in the obtained auxiliary rectangle $CMDK$, the opposite sides are equal, that is, $MC = DK$ and $MD = CK$. Observing that the altitude dropped to the base of an isosceles triangle is its median at the same time, we see that $AM = MC$ and $CK = KB$.

The area of a triangle equals half the product of its altitude by the base to which it is dropped.

So, the area of $ADC$, $S_{\triangle ADC} = \frac{1}{2} \cdot AC \cdot MD = MC \cdot MD$;

The area of $CDB$, $S_{\triangle CDB} = \frac{1}{2} \cdot CB \cdot DK = CK \cdot DK = MD \cdot MC$.

Comparing the last two equalities, we conclude that indeed, $S_{\triangle ADC} = S_{\triangle CDB}$. Therefore, our problem is solved by cutting the triangle $ABC$ across the $CD$ line.

**Problem 36.**

Solution 1

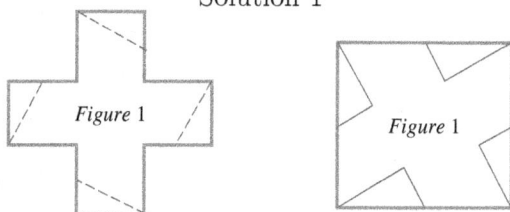

Picture 1

In the first solution, we cut four congruent right triangles from the cross, took Figure 1 obtained in the middle, and rearranged the right triangles to assemble the square as shown in Picture 1 on the right.

In the second solution, the problem is solved by drawing only two cutting lines on the cross (see on the left in Picture 2). Figures 1, 2, 3, and 4 are assembled to form a square as shown on the right in Picture 2.

Solution 2

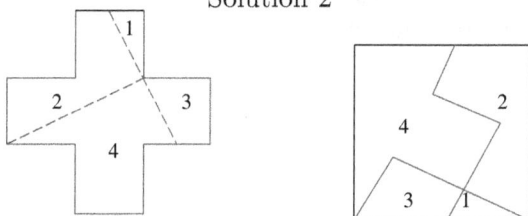

Picture 2

**Problem 37.** The statement of the problem is correct.

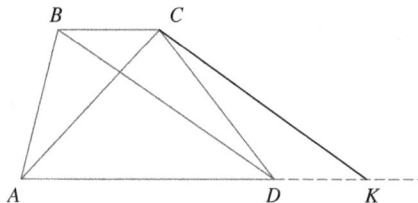

We need to compare the sum of the diagonals of the trapezoid $ABCD$, $AC + BD$, with the sum of its bases, $AD + BC$. To solve the problem, we will draw the line parallel to $BD$ at $C$ till its intersection at $K$ with the extension of $AD$, $CK \parallel BD$. It is given that $AD \parallel BC$ as the bases of the trapezoid, then $DK \parallel BC$ as well. So, we see that the opposite sides of $DBCK$ are parallel. Hence, this quadrilateral is a parallelogram. Therefore, $DK = BC$ and $CK = BD$. Now, we will consider the auxiliary triangle $ACK$. Two sides of this triangle equal the diagonals of the given trapezoid ($AC$ and $CK = BD$) and the third side is the sum of the trapezoid's bases, that is, $AK = AD + DK = AD + BC$.

According to the Triangle Inequality theorem, the sum of the lengths of any two sides of a triangle is greater than the length of the third side. Therefore, we conclude that indeed, $AC + BD = AC + CK > AK = AD + BC$.

**Problem 38.** We have to locate $M$ and $K$ on $CD$ and $AD$, respectively, so $MD = \frac{1}{3}DC$ and $KD = \frac{1}{3}AD$. Let's justify this.

Draw the diagonal $BD$. It divides the given parallelogram into the two triangles $ABD$ and $BCD$ of equal areas. It is easy to see that the area of $\triangle BKD$ equals one-third of the area of $\triangle ABD$. Indeed, the area of a triangle equals one-half the product of the base by the altitude dropped to that base. Draw $BP \perp AD$.

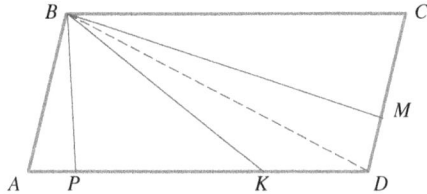

The area of the triangle $ABD$ is calculated as $S_{\triangle ABD} = \frac{1}{2} \cdot AD \cdot BP$.

Observe now that the area of the triangle $BKD$ is calculated as

$$S_{\triangle BKD} = \frac{1}{2} \cdot KD \cdot BP = \frac{1}{2} \cdot \left(\frac{1}{3}AD\right) \cdot BP = \frac{1}{3} \cdot \left(\frac{1}{2} \cdot AD \cdot BP\right)$$

$$= \frac{1}{3} \cdot S_{\triangle ABD}.$$

In a similar way, we get that $S_{\triangle BMD} = \frac{1}{3} \cdot S_{\triangle BDC}$. Therefore,

$$S_{\triangle BMDK} = S_{\triangle BMD} + S_{\triangle BKD} = \frac{1}{3} \cdot S_{\triangle BDC} + \frac{1}{3} \cdot S_{\triangle ABD}$$

$$= \frac{1}{3} \cdot (S_{\triangle BDC} + S_{\triangle ABD}) = \frac{1}{3} S_{\triangle ABCD}.$$

**Problem 39.** We draw three additional diagonals (see the broken lines) and as a result we will get six pairs of congruent triangles. The same digits indicate the congruent triangles on the picture.

The areas of the shaded and white congruent triangles will be the same. The only one triangle which does not have the congruent pair is the shaded triangle at the bottom of our regular 9-gon. Therefore, the sum of the areas of the shaded triangles is greater than the sum of the areas of the white triangles.

We suggest the readers investigate the rigorous proof of the fact that the indicated pairs of the shaded and white triangles are indeed congruent (never rely on the picture alone!).

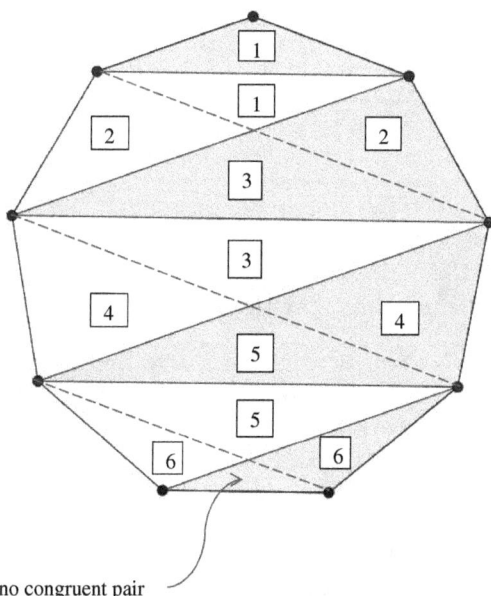

Triangle with no congruent pair

**Problem 40.** This cute problem-joke has a very simple solution. Bob will hear his friend first because the distance from his ears to

Steve's mouth is shorter than the distance from his mouth to Steve's ears.

**Problem 41.** Yes, he can, if he will repeat each step of both of his opponents in each game. If either of them won, the second would lose, and then Matthew would collect a point. In case of a draw, he would get two halves, which add to a point as well.

**Problem 42.** The tree which was the 20th for Alex was the 7th for Bryan; so, obviously, Bryan started his count from Alex's 14th tree. Bryan counted a total of 94 trees. His last tree corresponds to Alex's 7th tree. To determine the number of trees in this alley, Bryan has to add to his count another six trees located between Alex's 14th tree and 7th tree.

The total number of trees then equals $94 + 6 = 100$ trees.

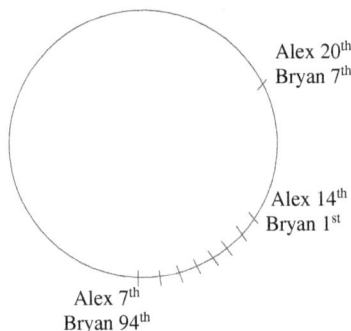

Alex 20$^{th}$
Bryan 7$^{th}$

Alex 14$^{th}$
Bryan 1$^{st}$

Alex 7$^{th}$
Bryan 94$^{th}$

**Problem 43.** $\frac{1}{3} + \frac{1}{4} + \frac{1}{5} + \frac{1}{6} + \frac{1}{7} = \frac{459}{420} > 1$. It is impossible, unless he is doing some things at the same time.

**Problem 44.** Yes, it is correct. Let there be $n$ people at a party. Clearly, each person has at least 0 or $n - 1$ acquientances. If we assume that every person has a different number of acquaintances, we will arrive at a contradiction. Indeed, if every person at this party has different number of acquaintances then

the first person knows no one,
the second person knows one person,
the third person knows two persons,
...............................
The last person has $n - 1$ acquientances.

But, this means that the last person knows everybody who is present at the party, including the first person, who knows nobody according to our assumption.

**Problem 45.** The lake will be covered in 9 days. Last year, it took a day to go from one lily-flower to two flowers. Starting from the second day, the process in both years is the same.

**Problem 46.**

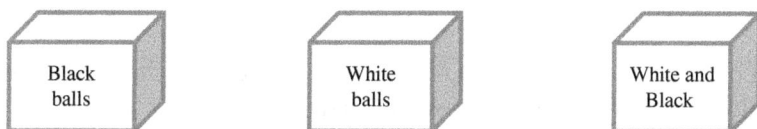

We need to take a random ball from the last box labeled "White and Black". If the taken ball is white, it implies that there are two white balls in that box (this is the only viable option because all the labels are messed up). The box labeled "Black balls" has to have either two white balls (which is impossible, we already identified the box with two white balls) or white and black (the only feasible option). So, the box labeled "Black balls" must have white and black balls. Therefore, the remaining box labeled "White balls" should have two black balls.

If the ball taken from the "White and Black" labeled box happens to be black, all the reasoning will be the same, and again, we will be able to correctly identify the balls in each of the boxes.

**Problem 47.** Denote by $x$ the first even Saturday in a month. The next even Saturday will be in two weeks; so, it will fall on $(x+14)$th. The third even Saturday in that month will fall on $(x+28)$th. The maximum days in any month is 31; thus, $x + 28 \leq 31$. It implies that $x \leq 3$, and knowing that it has to be an even number, we conclude that $x = 2$. Therefore, the last (the third) even Saturday in that month falls on the 30th. Respectively, the 28th of the month is Thursday.

**Problem 48.** We have to arrange the glasses taking turns — full, empty, full, empty, full, empty. And, we are allowed to touch only one glass. This will be accomplished if the second and the fifth glasses switch places. Now, we have to properly interpret the condition allowing "touching" only one glass. Clearly, we can't move the glass; it will not help to arrange the glasses as required. The only available option is to take the second glass with the water and pour over the water into the fifth glass.

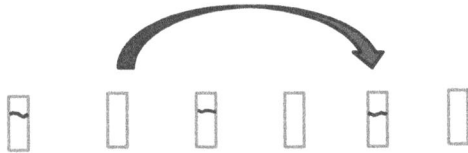

The problem is solved by "touching" only one of the given glasses because the fifth glass was not touched.

**Problem 49.** First, we simultaneously light one piece of fuse at the both ends and the second piece at one end. The first piece will burn in 30 seconds because it takes 1 minute for it to burn if lighted from just one end. As the first fuse finishes burning, we have to light the second fuse at its other end. It will take 15 seconds for the second fuse to finish burning. We will be able then to light the green tree in exactly 45 seconds, as was planned.

**Problem 50.** You need to turn on one of the switches and wait several minutes before turning it off. Turn on then another switch and enter the room with bulbs. The bulb that is on is connected to the switch that is on now. One of the other two bulbs should be warm compared to the other one because by keeping the first switch on, we allowed this bulb to be on for a several minutes, so it has to have a higher temperature than the third bulb. Thus, the warm bulb is connected to the first switch, and the third bulb is connected to the remaining switch. It turned out that a combination of some physics knowledge with a little bit of logic produces a good result.

**Problem 51.** If he said this statement on January 1st of the year 2020 (to make it vivid and easy to explain, let's pick a specific year), then yes, it is possible. In that case, the day before yesterday was December 30th of the previous year, 2019. If his birthday is on December 31st, then he was 10 years old on December 30th of 2019. He turned 11 the next day, on December 31st. Therefore, he will turn 12 on December 31st of 2020 and clearly, next year, in 2021, he will be 13 years old.

**Problem 52.** One can cover the whole distance riding on the bicycle for the same time as one would be walking half of the way between the cities. The motorcycle's speed is irrelevant for our decision (one has to spend some extra time riding the motorcycle anyway). Therefore, the fastest between the two alternatives is to ride on the bicycle.

**Problem 53.** Assume there were $x$ candies on the table, and several (irrelevant how many) boys took $y$ candies. Then, there were $(x - y)$ remaining candies available for the other boys. The next boy took $\left(\frac{x-y}{10} + \frac{y}{10}\right)$ candies. This sum is simplified to $\frac{x-y}{10} + \frac{y}{10} = \frac{x}{10}$. We see that this boy got $\frac{1}{10}$ of all candies available at the very beginning. This implies that every boy (no matter in which order) got the same number of candies, which is $\frac{1}{10}$ of all available candies. Since we know that all the candies eventually have been taken, then there were 10 boys, each of which took one candy.

**Problem 54.** First, we have to emphasize a piece of very important information given in the problem which can easily be overlooked. The moment the professor was interrupted, he said that as the result of arithmetic operations we have to get a number representing some year. So, we may conclude that it must be a natural number. Let's keep this thought in mind and introduce now a few variables. Assume the professor was born in $x$ year. Then,

$$\frac{(x + 43)(x + 45)}{x} = n \text{ and } n \in N.$$

This expression can be modified as

$$\frac{(x + 43)(x + 45)}{x} = \frac{x^2 + 88x + 1935}{x} = x + 88 + \frac{1935}{x}.$$

As we observed before, the last sum is a natural number. It is possible only when $x$ is a divisor of 1935. Logically, because $x$ represents the year in which the professor was born, this number can only be 1935 itself. So, the student was correct in his statement that there was enough information for him to be able to figure out the professor's age.

**Problem 55.** Each character represents a mirror image of digits from 1 to 6 in a reflection through a vertical line. The number following 6 has to be 7.

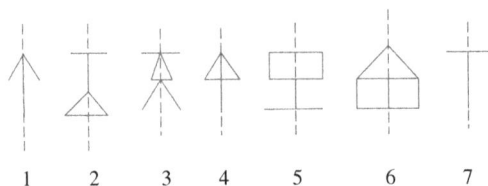

$$1 \quad 2 \quad 3 \quad 4 \quad 5 \quad 6 \quad 7$$

## Final Problem — Moral Test

Let's recall the options:

You can save the old lady, who most likely would not be able to survive this night without your help; or, you can pay back your friend for saving your life and take him in your car; or you can pick the dream of your life (there is no way to know whether you come upon a second chance like this).

Throughout the book, we faced various difficult problems and evidenced that to find the path from a logical labyrinth, it is instrumental to clearly understand a problem's conditions and the question asked, i.e., the goal to be achieved. Do we want to save the old lady? Sure. Do we want to pay back the old friend and not leave him at the bus station in this terrible weather? Yes, absolutely. What about the opportunity to be with your second half for the rest of your life; should you take it? No doubt, you have to. How can you accomplish all three of these goals having a two-seater car? Clearly, you can accommodate only one person in the passenger seat in your car. What about the other two? You don't have to! Non-trivial problems require non-trivial outside-the-box thinking!

Here it is the solution:

> *I would give the keys of my car to my old friend and ask him to take the old lady to the hospital, while staying at the bus station with the person who I fell in love with from the moment I saw her.*

# Bibliography

Stephen Barr, *Miscellany of Puzzles Mathematical and Otherwise*, Ty Crowell Co, 1965.

BBC News, *Special Report: 1997: Chernobyl: Containing Chernobyl?*, 21 November 1997.

Barry R. Clarke, *Challenging Logic Puzzles*, Sterling, 2003.

Haskell B. Curry, *"The paradox of Kleene and Rosser"*, Transactions of the American Mathematical Society 50(3) (1941): 454–516.

Martin Gardner, *Aha! Insight*, Scientific American, 1978.

Martin Gardner, *My Best Mathematical and Logic Puzzles*, Dover Publications, 1994.

Martin Gardner, *The Colossal Book of Mathematics: Classic Puzzles, Paradoxes, and Problems*, 1st edition, W. W. Norton & Company, 2001.

Copeland Jack, *"Alan Turing: The codebreaker who saved millions of lives"*, *BBC News Technology*. 18 June 2012.

S. Jelenski, *Śladami Pitagorasa — Rozrywki matematyczne*, Poznań: Drukarnia Św. Wojciecha (in Polish).

A. Kirilov, "Construction program", *Quantum*, March/April 1996.

Boris A. Kordemsky, *The Moscow Puzzles*, Dover Publications, 2016.

Livio Mario, *Is God a Mathematician?*, Simon & Schuster, 2010.

*"Mathematical Puzzles of Sam Loyd"* selected and edited by Martin Gardner, Dover Publications, 2017.

D. Nyamsuren, *Quantum*, March/April 1994. Problem M109, p. 31.

George Polya, *How To Solve It*, Princeton University Press, 1973.

B. Pritsker, *Geometrical Kaleidoscope*, Dover Publications, 2017.

B. Pritsker, *The Equations World*, Dover Publications, 2019.

V. Proizvolov, *Quantum*, November/December 1994. Problem B130, p. 9.

V. Proizvolov, *Quantum*, September/October 1992. Problem B62, p. 31.

Alan H. Schoenfeld, *Mathematical Problem Solving*, Academic Press, Inc. 1985.

I. F. Sharygin, "So, what's wrong?", *Quantum*, July/August 1998.

A. Spivak, *Quantum*, September/October 1997. Problem B213, p. 13.

S. L. Tabachnikov, "Errors in geometrical proofs", *Quantum*, November/December 1998.

A. Yaglom and I. Yaglom, *Challenging Mathematical Problems with Elementary Solutions* (translated by J. McCawley), San Francisco, Holden-Day, 1957.

A. Yegorov, "What you add is what you take", *Quantum*, November/December 1994.

В. Болтянский, Ю. Сидоров, М. Шабунин, *Лекции и задачи по элементарной математике,* Москва, Наука, 1972 (in Russian).

В. Гусев, В. Литвиненко, А.Мордкович, *Практикум по решению математических задач,* Москва, Просвещение, 1985 (in Russian).

В. Михайловський, М. Ядренко, Г. Призва, В. Вишенський, *Збірник задач республіканських математичних олімпіад,* Київ, Вища школа, 1979 (in Ukrainian).

В. Чистяков, *Старинные задачи по элементарной математике,* Минск Вышэйшая школа, 1978 (in Russian).

Г. Гальперин, А. Толпыго, *Московские Математические Олимпиады,* Москва, Просвещение, 1986 (in Russian).

Г. Філіпповський, *Математичні Пригоди Слоненяти Лу і його друзів,* Київ, Грот, 2002 (in Ukrainian).

Д. Клименченко, "Математичні Софізми", *У Світі Математики,* Київ, Радянська Школа, 1985 (in Ukrainian).

Е. Игнатьев, *В Царстве Смекалки,* Москва, Наука, 1984 (in Russian).

И. Акулич, *"Квант" для младших школьников,* Квант, НПП "Бюро Квантум" РАН, #3 1999 (in Russian).

И. Акулич, *"Квант" для младших школьников,* Квант, НПП "Бюро Квантум" РАН, #3 2001 (in Russian).

И.С. Петраков, *Математические кружки в 8-10 классах,* Москва, Просвещение, 1987 (in Russian).

М. И. Сканави, *Сборник конкурсных задач по математике для поступающих во втузы,* Москва, Высшая Школа, 1980 (in Russian).

Н. Васильев, В. Гутенмахер, Ж. Раббот, А. Тоом, *Заочные Математические Олимпиады, Москва,* Наука, 1986 (in Russian).

С. Олехник, Ю.Нестеренко, М. Потапов, *Старинные Занимательные Задачи, Москва,* Наука, 1985 (in Russian).

Ш. Горделадзе, М. Кухарчук, Ф. Яремчук, *Збірник Конкурсних Задач з Математики,* Київ, Вища Школа, 1988 (in Ukrainian).

Э. Готман, З. Скопец, *Задача одна — решения разные,* Киев, Радянська Школа, 1988 (in Russian).

Я. Перельман, *Занимательная Алгебра. Занимательная Геометрия*, Москва АСТ, 1999 (in Russian).

Я. Перельман, *Задачи и Головоломки*, АСТ Москва, 2008 (in Russian).

Я. Суконник, *Математические задачи повышеной трудности*, Киев, Радянська Школа, 1985 (in Russian).

# Index

*Index*

www.ingramcontent.com/pod-product-compliance
Lightning Source LLC
Chambersburg PA
CBHW061235220326
41599CB00028B/5438